The amazing true story of how
one man developed a town
and then sold it to the world

BUFORD
POP 1
ELEV 8000

DON SAMMONS

Former Mayor of Buford, Wyoming

BUFORD One
Don Sammons

Published by August Moon Publishing
1712 Pioneer Avenue, Suite 100
Cheyenne, Wyoming 82001
AugustMoonAMP@cs.com

Library of Congress cataloging-in-publication data
ISBN: 978-0-9910715-0-0

Cover design and book packaging by:
W. B. Freeman Concepts, Inc.
P.O. Box 702052
Tulsa, OK 74170
inquiries@wbfreeman.com

Printed in the United States of America

TO

E.B.

with great appreciation for
inspiring me to write this book

CONTENTS

BUFORD
One

1

SOLD!

A DREAM FULFILLED

April 5, 2012, began quietly, like most days in my last thirty years. I awoke early, checked my e-mail and had breakfast. Then I put on a black suit and drove seventy-five miles to Buford, Wyoming, for one last time to see who would buy my town and for how much.

Yes, my town.

Sole owner.

Population 1.

And yes, my town was for sale on that day—in true Wild West fashion at *high noon*, and in truly modern terms, by Internet auction!

The day may have started in a normal way but it was anything but a normal day or a normal event.

I awoke that morning with giant question marks hanging in the air—all of which made for a degree of anxiety.

Would anybody actually show up to do the bidding?

Would this be the day my "town" actually sold?

If so, what would be the price? Would the minimum be reached? Would the amount skyrocket to astronomical levels?

Who would be the buyer?

I arrived in Buford that morning at nine o'clock to get the

coffee pot set up, and to get the place ready for the auctioneer and the media trucks. I also needed to move the tow trucks and snowplow out onto the parking lot so the people who might bid on them could examine the equipment. I met with the Williams & Williams realtor representatives who were handling the auction and sale.

I knew this day my life would change.

But I didn't know exactly how.

I had been told that potential buyers from eighty-four nations had contacted Williams & Williams about buying Buford. Even so, I faced the *potential* that I might be all dressed up for no reason. I knew I *might* not have a sale and at the end of the day, I'd sweep the floor, lock the door, and drive back to Windsor and begin to regroup for another summer of running the store.

Without a doubt, I could not *possibly* have anticipated how things would turn out.

2

THE JOURNEY

WHY ME AND WHY BUFORD

Life happens in process. I have no doubt about that. It is a process that brought you to this moment of reading my story. It was a process that took me to Buford. It was a process that led me to sell it. In the midst of process, a story develops—not always by design at the outset, but over time, loaded with design. At least, that's been my experience. It is probably your experience as well.

How I came to be in Buford in the first place is not a complicated tale or a disjointed process. It was a story with several important segments to it—a story that took my first thirty years to live. When I look back, I can see that each segment of my story had an important element that came into full play once I sought to buy the Buford property.

Like most people, however, I lived in the moment and went with the flow—there wasn't a lot of time or energy for gaining perspective or for mapping out details in advance.

If you are asking, Why me?—as many people through the years have asked, my response would probably be, Why *not* me? And in that, there's a question, Why not YOU? The Buford store could happen to anyone.

As Normal as Apple Pie and the Wild West

I grew up in St. Louis, in a pretty normal family. Like most young boys in the fifties, I played cowboys and Indians. I liked to watch "Westerns" on television, and I daydreamed of life in the Wild West. The "West" to me was a place of wide-open spaces, lots of adventure, and the opportunity truly to live an independent life. All of which was . . . and still is . . . appealing!

I wasn't very interested in school—the world outside the classroom was much more appealing to me, and especially the world beyond my neighborhood and city.

When I was in high school, and before I finished high school, I announced to my parents that I wanted to leave home. They were opposed to the idea, of course. My plan was to drive with a friend to his grandparents' home on the West Coast and to live by the beach. What teenager in the 1960s didn't dream of that—at least for a brief moment!

The more my parents resisted the idea, the more I wanted to go. And finally, my father had a talk with my uncle, who convinced him that the car in which we were planning to make the trip probably wouldn't make it very far outside St. Louis. He also persuaded my parents to see the wisdom of allowing me to go, with the reasoning that if they allowed me to go, then the decision for me to return would be my decision—and conversely, if they refused to let me go, the more I would want to go and finally would find a way to go. So, they reluctantly agreed and off we went.

To *their* surprise, and my uncle's surprise, the car did *not* break down and we made the entire trip to Newport Beach without incident.

To *my* surprise, my friend's grandparents were happy to see him and were willing to take him in and provide a home for him—but not for me. I wasn't family, and they quickly made it clear that they expected me to find a job, pay my own way in life, and further, to live on my own, not in their home.

I was there almost a year, working in various odd jobs, and then, I made a decision that I wanted to pursue a career in telephone communications. I reasoned that everybody would always need a phone, right? And phone equipment would always need to be installed, repaired, and upgraded. I did just enough research to determine that the quickest way to get communications equipment training, and to be paid for receiving the training, was through the United States military.

I had another persuasion exercise in front of me. I was only seventeen at the time that I made this decision and that was too young to enlist in the Army without my father's permission. So once again I made my appeal, and once again, my father agreed. I think he saw a degree of responsibility in my choice, and even though the Vietnam conflict was escalating, the draft hadn't yet been instituted and by enlisting, I could choose a career—more or less.

Frankly, I wasn't opposed to the war. I was a real patriot in my heart—a lover of all things American. I saw honor in serving my country with military service. And I didn't hesitate to sign up. I was assigned to be a "radio operator." Before I knew it, I was sent to Germany.

Much of my time was spent playing variations of "war

games"—most of which involved groveling around in very cold German mud doing things that made no sense to me but which seemed to be supremely important to my commanding officer. Keep in mind that I was seventeen years old. I considered my options and decided after a year, if I was going to be out in the mud, I'd rather be in *warm* mud. So I volunteered to extend my enlistment in the military and serve in Vietnam.

I was granted a transfer.

Veering "Off Road" to Vietnam

Vietnam was not the horrible experience for me that it was for many servicemen, at least not physically horrible. Those of us who were radio operators tended to be in facilities somewhat distant from the fighting and also a little remote from the hierarchy of a base—all of which made it easier for us to do our work and not be detected, and therefore, not be destroyed. I was naïve in lots of ways, and was a bit surprised at how easy it was for American military personnel to get drugs and other things considered either illegal or immoral back in the States. I didn't have much desire for that, but I knew the protocol—a carton of cigarettes and a ten-dollar bill could be exchanged for a look-alike carton, but with each of the twenty packs filled with marijuana joints or a little "speed." It was a way to escape the humdrum routine and to cope with the body bags.

As part of a communications team, I also discovered that I didn't need to follow all the rules as strictly as those who were carrying guns—I could wear my hair a little longer, and didn't have the regular bunkhouse inspections. Mostly I was bored. And a little confused at the chaos I saw all around me. Part of the confusion, of course, involved a sense that we had plenty of power and might to destroy everything and anything we might want to destroy in that little nation, and yet we chose not to do so. That made no sense to me, politically or in the military world.

After I completed my active-duty deployment in Vietnam, I returned to the U.S.A. and was assigned to a base in Missouri. By that time, I had been out of the country for several years and coming "home" was a little bit of a culture shock. I didn't know where the military might send me next. I didn't feel as if I really fit into life back in Missouri, so while I was "home" and somewhat had my weekends free, I decided to enjoy life as much as possible. For me, that included motorcycles. A buddy offered to loan me his motorcycle one weekend and I eagerly took him up on the offer. I thoroughly enjoyed the freedom of bike riding and could hardly wait to hit the open road!

A Different Sort of "Wounding"

I hadn't counted on the driver of a large Chevrolet truck failing to see me as he pulled out of a parking lot. He hit me broadside, catapulting me into the oncoming lane of traffic. Although I was wearing a helmet, the impact was so great it knocked the helmet off my head. Those who saw the accident were shocked that I survived, and especially when they saw a vehicle run over me. I was grateful that vehicle didn't run over me any higher than my leg.

I had managed to make it through Vietnam without a wound, but as that day drew to a close, I found myself in a hospital with a seriously crushed leg and one of my heels hanging from the rest of my foot. The surgeons managed to stitch the heel back to my leg, but in the process of being at the county hospital and then transferred back to a military hospital, two things happened—an infection set into my wounded leg, and the orthopedic surgeons who needed to reset the broken bones in my leg were unable to do their work for several months. Even after the first couple of surgeries, the infection lingered and kept the bones from knitting. I seemed to be on an endless cycle of infection and bone surgery, all of which were compounded by very strange military bureaucratic decisions—the military couldn't seem to decide if I was retired or on active duty.

If I had been bored in Vietnam—well, I almost went stir crazy in the hospital.

Month after month passed, and when the necessary surgeries were completed and my wounds had healed enough for me to go to rehab in the hospital annex, I discovered that my wounded leg was a couple inches shorter than the other. I refer to those months as ones in a rehab hospital, but there was very little rehabilitation going on. I was in a long leg cast that was eventually shortened to a below-the-knee cast. By that time my leg was so stiff that the most rehabilitative maneuver I could do was have a nurse or aide help me dangle my leg over the side of the bed and then have someone pull down on my leg until I couldn't stand the pain anymore. Eventually I was allowed to go home on weekends, and I found that I could make it around my parents' house on crutches. Cast and crutches were discouraging to me—being incapacitated was *not* a way to live, and especially if

I had no recourse. For the first time in many years, I felt myself without choices, and for a guy who had dropped out of high school, been through Vietnam, and was ready for adventure, CHOICE was extremely important!

The silver lining in all this was that my term of enlistment in the United States Army was nearly over by the time I had been in the hospital and rehab center for a total of a year. Rather than require me to complete the month or so of my enlistment, the military gave me an honorable discharge and suddenly, I was a civilian again—back in St. Louis.

It had been a long way around the world to return to the city of my birth.

I was barely twenty-one.

So . . . NOW What Do I Do with My Life?

I was glad to be out of the hospital more than I was glad to be out of the military. But I also realized that I had not gained from the military what I had originally sought. I was *not* trained in the installation and repair of communications equipment, and I really was no closer to becoming a telephone repairman than I had been four years earlier. I took a look at the overall job scene and decided that climbing poles might not be the best job for a guy still recovering his full strength with one shorter-than-the-other leg, so I tried my hand at a couple of other career options.

I took a couple of sales jobs—even a stint at selling cars—and then I settled into being a fireman. The good news about a firefighter's job was the schedule—I would work several days around the clock, and then have a few days off. Not one to sit around idle, I discovered the world of moving vans to occupy me on the days away from the firehouse.

A friend began to hire me to help load his van—in moving terms, "lumping." I quickly saw the job as part physical labor—which felt good to do and good to be *able* to do—and part puzzle solving. I took it as a personal challenge to get as many things into the back of his moving van as possible. It was a mindset, and a skill set, that would serve me well later.

A little known fact to the general population—a moving-van driver gets paid for the *weight* of a load and if more can be loaded through careful strategic placement of items, the better the pay!

One day I asked my friend if I could go with him on a haul. He agreed—perhaps glad for the company, perhaps glad to know I'd be there to help him unload. I already had a chauffeur's license but had never driven a big rig before. I enjoyed the trip, and it was on that trip that it became crystal clear that the *really* good money was not in loading a van, but in driving a van.

Over the course of several trips with him as my "role model teacher" about all things related to driving a moving van, I made the decision to seek work with a reputable national moving-van company, driving my own rig. My friend was a little disappointed to lose me as a loader, but he also recognized that I needed to be doing more with my life, and in general, he was supportive of my decision. By the time I went to the moving-van company to work out an employment deal, I knew the jargon of the moving-van world and I knew how to drive a rig. I was hired on the spot and tossed a set of keys. The tractor I was assigned had already been scheduled for a haul to California. After a few minutes of learning some key differences between the Freightliner tractor I *had* been driving and the International tractor that I *would* be driving, I was off!

Major Decisions and Transitions

During these years, I also married. It was a decision and action that I wasn't really prepared to "live out" after Vietnam.

I take total responsibility for the demise of my first marriage, which occurred shortly after I recovered from my motorcycle accident. The gal I married was and is a fine woman. It's a shame she had to hook up with me at that stage of my life. I wasn't at all emotionally ready to be husband material. I was still trying to adjust to life back in the United States and to rediscover—or perhaps *discover*—who I was and what I wanted out of life. I was drinking fairly heavily and wasn't at all inclined to give up the party life and settle down.

We were together for about eight years, and in retrospect, I find that amazing.

If I wasn't ready to be a good husband, I certainly wasn't ready to be a father. When my wife got pregnant, I left. I truly felt it was better for her to raise a child *without* me, than to try to raise a child *with* me.

I hasten to add that I did support our son financially all the way through college, and occasionally, I had brief encounters with him when I went to St. Louis to visit family and friends. But I was not an active "Dad" to him in any emotional or personal way. I regret that now.

Let me also hasten to fast-forward and add that we are in touch now, and I am grateful for his forgiveness and understanding. We are doing what we can to develop a relationship that can work between two adults. He is in his mid-thirties. He's a great young man, highly intelligent, skilled in his work, and responsible. I admire the man he has become and the way that his mother raised him. I'm just sorry that I can't lay any claim whatsoever to being a positive influence in his early years.

I made a second major decision toward the end of my twenties that turned out better. But not without a rocky start.

Once my wife and I divorced, I was free to pursue the moving business full-time, and I did. I enjoyed the trips. I enjoyed the challenge of loading a truck to the max, and of working as many days a month as I could work. And when I wasn't working, I was determined to party as hard as I worked.

One night I went out drinking, and unfortunately, I went out on the town in the moving van I had been assigned. I went to a bar I frequented fairly regularly, and when I left that night, I sideswiped three or four vehicles as I pulled out of the bar parking lot. I didn't stop, but the incident was witnessed by someone. That person called the moving company and fortunately for me, the person who answered the call "took care" of the incident.

She told me, "This is your one favor. Get your life together."

I thought long and hard about what she said and decided, *I'm going to take thirty days off and go into rehab.* And I did.

The therapists there told me that they weren't sure if I

had a "drinking problem"—in other words, whether I was an alcoholic—but we all agreed that my drinking was creating problems! I, and they, chose to err on the side that I *might* have a problem, and to address it.

Throughout my life, I have had a very good experience in simply making up my mind to do something, and then be able to follow through on the decision. When I returned from Vietnam, I made a decision that I would never use drugs again—not with anybody or for any reason. And I never did.

Later, I gave up cigarettes. I actually gave them up three times—the first "quit" lasted six months, the second "quit" lasted three days, but the third "quit" was definitive. I haven't used nicotine for decades. It wasn't the threat of cancer, but rather emphysema and breathing problems—I couldn't take that risk living at the high altitude of Buford.

Rehab worked for me. I stopped drinking at that time. Permanently.

While I was in rehab, I met a gal named Terry who was also in recovery—for both drugs and alcohol. Being patients together in a recovery center is not exactly the ideal circumstance to find a girlfriend, but that's what eventually turned out.

After we were both out of rehab, we reconnected and she began to go with me on some of my longer hauls cross-country. We seemed to do well in the "moving business" together.

As a mover, I chose to make my home base in Newport Beach. I liked the weather there better than St. Louis, and I really didn't know any other place as well as I knew southern California. This time, of course, the relocation was on my terms and at a significantly higher pay grade than the gas-station job I had during my teen years.

Both Terry and I enjoyed the relaxed beach lifestyle of that area, and there certainly was enough business to keep me loading and unloading as many moving vans as I wanted. Eventually, I found myself in association with Allied Moving and stayed with them for a number of years.

A Mastery of MOVING!

I'm not at all bashful in stating that I was a MASTER at loading a truck and making the most of every square inch of space in it.

I made a key business decision that I would not only drive for Allied, but that I would OWN my entire rig—both tractor (the cab) and trailer (the van). Many drivers own their own tractor, but I did what few do. I bought my own trailer and customized it so I could haul much more per load. I made the well—the part between the wheels—deeper and broader. The result was that, most of the time, I could haul at least fifty percent more in my moving van than most movers, sometimes a hundred percent more. That, of course, meant that every load I hauled was worth more money to me since movers charge by the weight of a load.

I became very good at hauling the exotic and unusual.

Perhaps the most interesting thing I ever moved was a curio cabinet in which a woman had hung her grandfather's skeleton. Both her grandfather and father had been physicians, and her father had the grandfather's body turned into a skeleton for medical training purposes. I told her I wouldn't "pack" that item but I told her *how* she needed to pack the item for safe transport. I must admit that I was a bit startled when I first saw this curio cabinet in the corner of that woman's living room!

I once transported one of Houdini's original "escape" chests.

I certainly transported a number of custom-designed high-dollar automobiles.

I also became very good at loading my van far beyond the expectations of most of the people I moved.

One time I pulled up to an executive's home and he asked me bluntly, "Where's the other truck?" This man had watched several of his fellow executives pack up and move. He had concluded that they had *less* to move than he had, and each of these execs had required two moving trucks to relocate their possessions.

I said, "I'm the only truck that I know of."

He seemed disgusted. "Well, we need another truck. I have a lot more stuff than my colleagues who have already moved and they all needed two trucks, or at least part of another truck."

He added, "I have all the stuff in my house, my boat, and my Porsche! You need to call for another truck right away. I want to be out of here *today*."

I said, "Let me look around and if I don't think I can get it all, I'll call Allied right away." After I had looked through his house, I told him I didn't think there was a problem.

He repeated, "I'm not leaving anything here today." I nodded in full agreement.

I called for a flatbed tow truck to help me lift the boat and car. We loaded the boat and its trailer into the belly of my truck. I built a floor over that, after packing things all around the boat. Then we loaded the Porsche on top of the artificial floor and packed furniture and boxes all around that. And, I still had several cubic feet left over.

He was amazed.

I learned very quickly in the business that every bit of space MUST be filled with something, and that most items can be reduced to or packaged as squares and rectangles of some kind. Heavy items on the bottom, lampshades on top! That's a method that works ninety times out of a hundred.

For me, the loading of a truck was a giant puzzle, and I was very good at working that puzzle.

I didn't know it at the time, but that concept had a direct impact on the way I "loaded" my store at Buford.

Good Lessons in Customer Service

The moving business also afforded me good insights into customer service.

I learned very quickly to value the fact that I was working for people who were putting a big part of their "life" into my hands, and into my van. I knew more about moving than they knew, but I also knew that I was moving *beloved* possessions and *prized* purchases. The moving business was part skill, with a big percentage of customer service.

When it came to my running a store many years later, I discovered that the "formula" was very similar. I knew more about Buford and the items in the store than the customers knew, but I was also seeking to serve people—to provide for them items they would value through the months and years ahead, and to make their travel easier or more pleasant.

One of the things I did often was to show my moving customers the inside of my moving van. I kept the inside of my truck spotless, with all of the ropes tied neatly, and a place for everything and everything in its place. I presented myself as a professional, and I kept a professional attitude at all stages of a move. The more nervous a customer seemed to be about the moving of their possessions, the more important it was for me to show customers that I took pride in what I did, and that I was in the moving business because I *wanted* to be in the moving business helping people and taking good care of their possessions.

I routinely hired guys in various locations who had a professional attitude, worked hard, and were pleasant to be around. Through the years, I had people in most of the major cities who were eager to get a call from me to help out with a move. Developing that large team of "extras" was a great benefit to me.

What I never really understood in either the moving business or the retail store business was the way some people could switch gears in their attitude, words, and behavior when it came to taking on the roles of "boss" or "server."

People who could be very friendly on a personal basis could suddenly become a little tyrannical if they felt they were in charge or in control, and vice versa. They could become far less confident if they suddenly recognized that they were in a subordinate position. Now, I certainly understand different roles in life, and the fact that somebody needs to lead. But I don't understand treating people with different attitudes. I didn't understand it in the moving business, and I didn't understand it in retail.

People were sometimes very rude when it came to their telling *me* how to load my truck or handle their possessions. I certainly had no intent of being careless in the way I moved items or loaded them. I packed items with extreme care. And, trust me, I would have done that without the person looking over my shoulder or dictating my every action.

People in the store could be very rude at times in telling me what they thought of my gas prices. They had no clue, really, what they were talking about in many cases. I didn't need the hassle they interjected into my routine. They likely would have *hated* being treated the way they treated me if the store-owner shoes had been on their feet.

But in the end, I decided that human nature is what it is. There's no changing what other people say or do—not *really*. I could only control my attitude and responses. And long ago, I stopped trying to please everybody all of the time. When I stopped trying so hard, I seemed to please more people more of the time!

Constant Variety . . . But Still Restless

I enjoyed the constant variety in the work, but over time, I found being based in Newport Beach a little confining. There was plenty of work—at times, it seemed to me that everybody in the entire area was moving at least once every two years! But, I had relatively few long hauls to give me a change in scenery. I wasn't exactly bored, but I was a little restless.

After we had lived in Newport Beach for about two years, Terry and I took a trip to visit friends of mine who lived in Wyoming. The guy had been in automotive sales with me and his wife was from Cheyenne, so they had relocated there. I stopped by to see him any time I had a haul that took me through Cheyenne along I-80.

Terry and I both fell in love with the mountains and the challenge, adventure, and wide-open spaces of the state, and we decided to make it our home. We settled in Cheyenne initially and lived there for about a year while we looked for ranch property. We finally found what we were looking for in the Buford area.

We recognized that life would have different "difficulties" in Wyoming—every place has some degree of difficulty and Newport Beach was not immune to challenges. We felt up to taking on the challenges that Buford might throw at us.

Ninety-mile-an-hour winds? *Only ninety miles an hour?*

Difficult access to the highway in the winter? *There's got to be a way.*

No friends within a twenty-mile radius? *We'll make friends.*

No shopping malls close by? *We'll live off the land.*

We looked forward to having a big garden, riding horses, learning what it meant to have a limited herd of cattle, and what it might mean to "tough out" being snowbound.

We were ready for Wyoming!

3

THE RANCH

A LOG HOME AND A QUEST FOR ACCESSIBILITY

When Terry and I moved to Wyoming, we rented an apartment in Cheyenne, and almost immediately, began looking for property in the surrounding area. We weren't interested in city living—we wanted to be as close to the natural beauty of Wyoming as we could be.

We certainly had nothing against Cheyenne. It was and is a nice city—small, easy to navigate, friendly, with all of the professional services and culture a city should have.

But a few miles west is wide-open land and even bigger skies.

We scoured the entire region. We looked in Steamboat Springs, Fort Collins, Laramie, Estes Park, north of Cheyenne, and finally the area around Buford. We eventually settled on a ranch—initially forty acres but eventually eighty acres—about three miles south of I-80 and, as it turned out, just about three miles south of Buford.

Two state parks and a national forest are located near Buford. There are a few large longstanding ranches in the area as well. Overall, the wilderness quality of Wyoming has been well preserved.

On the property we eventually purchased, the owners had built a home that they had fully intended as a "retirement home." They lived in New York and had traveled to the Buford area, and had been captivated by the open spaces and mountains. While they were constructing their home, they went back and forth several times, but they never really stayed in Buford longer than a week or so at a time.

When they actually *moved* to Buford, they discovered that the wife could not tolerate the altitude over an extended time. She had terrible breathing problems, and that ended the "dream" of Buford for them. For our part, we had already experienced plenty of time to acclimate to the altitude and to the prevailing isolation of that part of Wyoming.

At the time I first saw the house, it had been on the market a couple of years. I hasten to add that I did NOT see the house in winter. And therein was part of the problem that had *kept* this lovely home on the market.

Snowfall on Our First Day at the Ranch!

We moved to the ranch on September 20, 1984. Just as I finished unpacking the truck, the snow began to fly. I thought it was "neat."

It may have been irony, or an omen perhaps, that our first day there was a snowy day. Mother Nature was giving us a preview of what was to come, and to come, and to come.

At the time, however, we thought the snow was magical and beautiful. It was only later that the full scope of the problem was revealed. Our biggest challenge was neither the weather nor the loneliness. It was *access.*

And accessibility was vital to me. Now, we had been told by the seller that the ranch had accessibility to I-80. But it is one thing to be three miles from a major interstate highway. It is another thing to be able to GET to that highway, in all types of weather, at any time of day. Although there was a gravel road leading from our log home to I-80, we discovered very quickly that we did not have real ACCESS. And for a guy in the moving business, and for a guy with a wife and baby son, access was critically important.

First awareness factor: the three miles to I-80 were "as the crow flies." The gravel road was thirteen miles from I-80, and it was only the benevolence of a neighbor ranch that got that distance down to a more reasonable number.

Second awareness factor: the road was *dirt.* Furthermore, the three miles was neither straight nor flat. Even with small amounts of snow, the drive could be slick and hazardous. It had numerous curves, dips and hills, and naturally, given the snow and wind, was prone to washouts and holes.

I had every intention of continuing my work in long-haul trucking for Allied Moving. In the moving industry, the corporate dispatchers really don't care where a trucker is located—at least, *precisely* located. An available trucker will be notified where a load needs to be picked up and delivered, and generally speaking, what size and type load is involved, and the trucker can do some quick calculations to determine if the load is profitable. At times, a "return load" or related load might be factored in. The good part about I-80 driving was that loads tended to be long haul, not just around the corner. The longer the drive, the better the profit. Generally speaking.

I should have recognized the accessibility problem on moving day when I couldn't get my own Allied rig down the road to the log cabin home. We had to rent a smaller truck to ferry our possessions to the ranch. But I remained undaunted.

I quickly decided that if I just put chains on my car—a Lincoln Continental Mark IV at the time—that I would have no problem. I was wrong. It is no easy task to get snow chains on a Lincoln, by the way.

I then decided that it might be easier to go across the fields in the snow, rather than stick to the roads. I was wrong.

One time we tried to get out "cross-country" and we came to a gate. The horses began to follow our lipstick-red Lincoln with its white top with a group of a half dozen horses following it. The lead horse began kicking at the car as it spun out in the snow, apparently thinking that the Lincoln was trying to take over its leadership role.

I frankly was surprised to see the horses. I hadn't realized that I was on private property—I thought I was on railroad right-of-way.

We managed to outrun the horses and reach the other side of the field without hitting a rock buried just under the snow—which was a major feat.

I then decided to trade in the Lincoln for a more Wyoming-worthy vehicle. I bought a Jeep. On the day I bought the Jeep, we finished the transaction just before dark. I wisely chose to stay in town rather than try to get out to the house in the pitch dark. When there's no moon or stars—which happens often in winter—there's no way a person can rely on headlights to anticipate how deep the snow might be. It might be just a few inches, or it might be snow that has filled a gulley or little ravine and is several feet of powder. It might be snow over hard-packed earth, or snow just barely covering the top of a jagged rock. There was one time I literally crawled the last two hundred yards to the log home so I could locate any deep holes and rocks under the snow.

One day Terry and I got to our gate about nine o'clock in the morning, which meant we still had three miles to go. By five o'clock that night, we could see the house in the distance but we still weren't home. We had spent all day digging out of one snowdrift after another.

Being in a Jeep hadn't made a bit of difference on *that* day!

I tried a snowmobile—and a snowmobile with a little trailer attached to it so Jonathan could go with us. That didn't work in the wind. The snow was blown off the ridges and we were riding on dirt much of the time.

I tried cross-country skis but that was slow going and ineffective.

At times, I literally walked out to the highway on snowshoes. They were even slower than skis.

I decided with firm conviction that the seller's idea of accessibility was not *my* idea of accessibility—at least not in the winter months.

We tried all sorts of maneuvers and vehicles and gear shifting. For *most* of a year, the three miles was "iffy."

One day during our second year at the ranch I was returning from a haul to California and in Grand Junction, Colorado, I came across a Tucker Sno-Cat that was for sale. Aha! This is what they use to groom the trails at the ski resorts. This one was thirty feet long, articulated or "hinged" at the halfway point, with four tracks each fifteen feet. It was totally enclosed and could carry eight people. It had plenty of room, no doubt about it.

The Sno-Cat had rubber cleats so it could be driven on pavement. The top speed was seventeen miles per hour—most of the time I drove it about ten to fifteen miles an hour. Speed wasn't important. Getting there, was.

I loaded up the Snow-Cat in the semi and called Terry to say, "I think I have our access problem solved." When she asked me what I had purchased, I said, "A snow machine."

She thought I had purchased a Trackster, which is what my neighbor Edwin had. It was a much smaller version of the Snow-Cat. I stopped in Cheyenne to get the oil and transmission fluid changed. Our rancher neighbor heard about the call and my stop for oil. He said, "If that's the case, he bought something a lot bigger than a Trackster."

Indeed.

But the Sno-Cat finally solved the problem!

The folks at the Lone Tree Junction Store allowed me to park my moving van there. In exchange for the favor, I bought my diesel or gasoline from them. They had the electric company put a meter out by the parking area so I could plug in my truck, which had engine heaters, fuel heaters, oil heaters, water heaters, and all other necessary heaters. That kept me from having any difficulty getting started. And while I was away, I could plug in the Sno-Cat, so I had no difficulty getting home.

Having the Sno-Cat made the commute seem "civilized" to me. If I was cross-country skiing or driving a snowmobile, I always had problems staying clean. Either the snow or mud was splattering, and the jumpsuit I was wearing had to be removed carefully and stored so I didn't get mud on the garment I was wearing for work or a visit to Cheyenne.

And, of course, I also learned the value of a shovel.

A Shovel in Every Vehicle

One of the best pieces of advice my ranching neighbor Edwin ever gave me was this: "Put a shovel in every vehicle you own." After a little pause he added, "You'll know you're doing the right thing if you ever need to use the shovel."

I learned that lesson the hard way one day when our son, Jonathan, was only about four or five years old. There was a snowdrift in a little gulley ahead of us and I decided that we could "jump" that drift in my new truck if I just got up enough speed.

Jonathan said a couple of times, "Don't do it, Dad."

I replied, "Aw come on, we can make it. It will be fun."

Well, we soared all right, but landed smack dab in the middle of the drift. Jonathan opened the window on his side of the truck and crawled out. "Where are you going?" I asked. He said, "To Edwin's. Come get me when you get this dug out." He scrambled up the other side of the ravine to Edwin's house about a hundred yards away.

I worked for hours with my shovel before I dug out of that drift and went to pick him up.

Edwin, of course, had decided I would learn my lesson if he did *not* come to help me.

Settling In on the Ranch

Edwin had also predicted that Terry and I would only survive at the ranch about six months before we moved. I'm glad I didn't know that at the time.

I was *determined* to beat Mother Nature and solve the problem of accessibility.

I loved the ranch lifestyle.

Our log home had 3,000 square feet. It was a ranch-style house that had two wood stoves and an electric forced-air furnace for heat.

I put up a nice entrance to the ranch—poles and a crosspiece with wagon wheels hanging from it. It was a ranch entrance worthy of Wyoming, at least in my opinion.

We had our own well and septic tank, and disposed of our own trash, mostly by burning it. We had to be self-sufficient when it came to those matters.

One day I decided to clear a hay field by burning off the stubble. For the record, that is NOT a good idea. Also for the record, I didn't burn down the house.

Over time I learned to "read the ridgelines." I studied the lay of the land. Usually, a ridgeline had less snow, or the wind had blown the snow away from the top of the ridgeline. It took longer to go cross-country following the ridgeline, and you needed to have permission from the rancher who owned the property to cross his land. Over time, I got all the permissions I needed.

I chopped wood so we wouldn't have to use the furnace. I'd buy wood from a log truck, which would come out to the ranch and drop off the thirty-foot logs for me to cut up into wood-burner-sized pieces. It was hard work, no doubt about it, even with a chainsaw.

I'd have about twenty cords of wood stacked up around the house, and that amount would usually last about two years.

Electricity was unpredictable and very expensive—it was provided by a co-op in the area. Phone service was also unpredictable, and of course, at that time and until the cell tower was located at Buford, there was no "mobile phone" service. The wires were laid on the ground and over creeks. About three days in any given week, there was no electrical or phone service.

A phone repairman told me one time that in the line that went from my house to the main highway, there was every kind of wire used since phone service was invented.

I found it a bit ironic that I was living in a wilderness environment with no phone service when I had desired at age seventeen to enter the military to become a phone-service expert! Resolving the transportation problem was easier than resolving the electricity and phone-service problems.

We bought some horses—now *this* was what ranching was all about!

Terry wasn't into the livestock as much as she was into gardening. I often teased her by calling her "Mother Earth." She was into natural nutrition, natural fabrics, and a "back-to-nature" lifestyle.

We quickly learned, as most other ranching residents in the area, to go to the stores in Cheyenne infrequently, and to stock up. We lived out of our pantry when we were snowbound.

We enjoyed watching the birds and wild animal species in the area—fox, lynx, mountain lion, bobcat, bear.

And every time we rode horses at the ranch, I found myself thinking, *THIS is for me!*

New Neighbors and Friends

Moving to Wyoming, and especially to the ranch, meant more than a change in lifestyle and scenery. It meant new friends and neighbors!

Buford was far more than a store along a highway. It was a never-ending living drama with a large cast of colorful characters.

Beverly and Richard were good friends. Richard died one day after having a massive heart attack while driving just a few miles from the store. He had stopped in that morning for a cup of coffee and we talked for a few minutes. I missed him every day after that.

Gladys and John had ranched in the area for many years. They were of the "old school" when it came to morals and values. They were like a "rock" in times of trouble.

Mark and Laura were both attorneys who lived in the Buford area. Mark was an attorney for Laramie County, and Laura had a private practice. After they had two sons, however, Laura gave up her law practice and stayed home to raise their children. And raise alpacas. Over the years, I watched their herd grow and I went to the celebrations that marked the annual sheering of the animals. Frankly, I knew NOTHING about alpacas before I met them. I certainly didn't know Wyoming was a place for raising alpacas.

Grizzly was a man who moved to the area about five years before I left Buford. He lived with his sister Marion on one of the "new ranches."

In the Buford area, some of the ranches established in the 1800s were parceled into thirty-five-acre spreads. Grizzly and Marion had purchased one of those.

His nickname was "Grizzly" because he had a long beard. He was an itinerant preacher. He told me that he had been struck by lightning numerous times. He seemed, nonetheless, like a sane person. On the other hand, he often did what seemed like highly irrational things.

One time, he waited until icy conditions set in, headed for gasoline, slid off the road, got out of his truck to see what he might be able to do, dropped his keys and cell phone, and was stuck there all day until a neighbor happened to see his truck off the road late in the afternoon when he was on his way home.

I advised him, "If you hear that a big storm is coming, drive down to the store and get your gas right away. Don't wait for the storm to hit!" He had a habit of waiting until the snow was really coming down before he headed for the store to get gasoline for his truck.

Now Grizzly had lived in Alaska for a while, and he replied, "Don, I *know* how to live here." Well, maybe not. I suspect he never encountered in Alaska the blizzard conditions that often enveloped the Buford area.

Edwin Davis. By far and away the most interesting of my neighbors and the closest one to me for decades was Edwin Davis.

Edwin was born in 1899. His parents settled in the Buford area, traveling there by covered wagon out of Boston. He showed me some of the trunks that were in that wagon, as well as mantel clocks that had been brought from the East.

Edwin was about five feet, three inches tall—and with rocks in his pocket, he probably weighed ninety-five pounds.

He had a very mechanical mind—he would have been a great engineer if he hadn't enjoyed ranching so much. He could build his own vehicles, devise his own systems, and even in his eighties, he managed to operate a cattle ranch with several hundred head of cattle. By himself.

He once went to an Air Force auction and purchased a large generator—he pulled it home on a two-wheel trailer. He claimed that the Air Force had never been able to get this particular generator running, and Edwin diagnosed the problem: "They put two magnetos on it." I have no clue what a magneto is, or if that is even a proper name for a legitimate generator part. Nevertheless, Edwin brought the generator to his ranch, took out the "magnetos" and installed a part of his own choosing, and the generator worked just fine!

He had purchased the generator as a backup system for his home. All of us had unpredictable electrical power in those days out on the ranch. Edwin decided the best place for the generator would be at the back of his house up a very steep hill. He had built a little shed on top of this hill and he rigged up a winch system to haul that generator to the top of the hill, removed the wheels, mounted it in the shed, and hooked it up to his house. All at age eighty-something, and by himself. I was amazed.

Shortly before I moved to the ranch in Buford, Edwin fell on ice while out feeding his cattle. He broke his leg. It took him all day to crawl his way back to his house and make a call for help. He was a tough ol' guy. No fat, all muscle. He was the epitome of a person I'd call "wiry."

Edwin had amazing stories to tell.

He liked to tell stories about Lucille. Edwin had two brothers. They were both married for a time to Lucille. She had her own reputation in the area as a hard-driven, no-nonsense woman.

He liked to tell me stories about his steam shovel that had been used in a quarry, of sorts, to provide the rocks used to create the bed for the railway.

He liked to tell about Tom Horn—the last man executed in Wyoming. Horn apparently had a good side and a bad side. You just had to be careful which side you were standing on! Cattle ranchers were trying to set him up in order to shut down his rustling behavior. The ranchers killed his son and pinned the blame on Tom. Tom was hung for the "murder." Edwin told me the sheriff came to his ranch when he was about twelve years old and told him that Tom Horn was not really hung—it was another guy who happened to be at the wrong place at the wrong time. Tom Horn was a big topic of speculation in the area when I first moved to Buford. He was part of the "lore" of the place that gave it a Wild West magic.

Edwin died at 105, asleep in his bed. He had no relatives and he definitely had people who knew he had money and they wanted it. As long as I knew Edwin, he had a housekeeper who cleaned up his place and cooked meals for him a few days a week. After his longstanding housekeeper retired, he hired a new housekeeper who was in her twenties. In addition to the few hours of housecleaning and laundry and cooking that she did each week, she wrote out checks for Edwin to pay his bills, and she eventually got power of attorney.

The day after Edwin died, she had his body cremated, which eliminated any possibility of an autopsy. The will in place at the time left everything to her and a little while later, she disappeared from the area. Nobody in the area even knew that Edwin had died until after his body had been cremated and the ashes scattered all over his ranch—which also meant that there was no memorial service for one of my favorite people in Buford.

Edwin often used the road in front of his place—which was actually a private road owned by the railroad. I used the road with the railroad's permission, and also Edwin's permission for the part of the road that ran alongside his property. When I sold the ranch, the new owners locked the gate and *nobody* had access.

Edwin and my wife were good buddies. He blamed me for her later leaving. He didn't know, of course, about her drinking problem. Edwin and I didn't retain the same depth of relationship we had before she moved out, but we certainly were cordial. Once Edwin had his mind made up, it was set in concrete. He would never have believed anything negative about my wife so I obviously had to be the fall guy for her "abandoning" the ranch life that included him. I understood that—I didn't like it, but I understood it.

In many ways, Edwin was my mentor when it came to survival in a generally "unsurvivable" place.

I know he would have had no understanding about my leaving Buford. He had absolutely no desire ever to leave his ranch. It was his *life* and he enjoyed it fully. He easily could have sold his ranch for millions of dollars and lived anywhere he wanted to live, in a style far more grand than his life on the ranch. But from *his* perspective, there was no better place to live, and no better way to live, than to live on his ranch and do the work of running a ranch.

It's hard for me to think about Buford and ranching without thinking about Edwin.

4

BUFORD

WHERE WIND AND WINTER MATTER

Most people have never heard of Buford, Wyoming.

With good reason!

It is in the middle of the proverbial nowhere, on I-80, the major highway linking New York City and San Francisco, about halfway between Cheyenne and Laramie. You might want to consult a road atlas at this point.

Yep, the middle of *nowhere.*

The greater truth is that hundreds of thousands of people pass by, or *through*, Buford in a given year. Tens of thousands of people stopped by the Buford Trading Post for gasoline, food, souvenirs, or various other sundries. Most people who have been there have a good memory of the place. And perhaps their overriding memory is: *Man, Buford is way out there.* Isolated is not an overstatement.

A Rich History for Buford

Although it may be remote, Buford has a rich history.

When Buford was founded in 1866 it was a military post called Fort Sanders. It was established to be a hub to protect the Union Pacific railway construction crews and the tracks involved in the Intercontinental Railroad as it moved west toward Promontory, Utah.

As the construction moved west, so did Fort Sanders. It obviously wasn't a wooden-walled fort. Fort Sanders was mostly a tent city, and it could therefore be highly portable and movable. Once the "fort" was positioned in Laramie, it became a more permanent structure, where its ruins can still be visited.

In its heyday, Buford had about two thousand residents, most of them foreign migrant workers who were building the railway.

In 1880 they put a post office at Fort Sanders for the permanent residents who lived in the area, and the post office was named the Buford Post Office in honor of Major General John Buford, who had been a hero in the Civil War.

There's a town outside Atlanta in Georgia that is also named Buford. That's where John Buford lived. He is credited with having fired the first shot at the Battle of Gettysburg. Although he was a southerner, he fought for the Union, which also made him noteworthy.

When the Mormons in Utah made a move to secede from the United States, General John Buford was reportedly the official sent by the U.S. federal government to negotiate with the Mormons and put the issue to rest. There is a likelihood, therefore, that he traveled the railway to get to Utah and he may have traveled through, perhaps even stopped temporarily, at the depot that bore his name, but there is no record that he ever visited there otherwise.

A County Seat but Never a City

Buford was established in 1866, Cheyenne the next year, and Laramie the year after that. That made Buford the second oldest major settlement in Wyoming.

For several years, Buford was the county seat for Laramie County. In the late 1800s, when the state was carved up into additional counties, land from Sweetwater County and Laramie County was taken to create Albany County. Buford became part of the new Albany County.

I had an old survey map in my office that showed Buford in Laramie County.

The fact is, however . . .

Buford was never an incorporated town. It was always a mail center, and later, a ZIP code—82052 to be exact. The post office boxes were located within or near whatever form of business or building was located at the site. That was the case from the beginning of postal service to Buford, and is now the case.

Cheyenne was a major hub for the railroad for many years.

The railroad built twenty-five steam engines *specifically* to climb over the mountain to Laramie. They were called "Big Boys." Only one of these custom-made engines remains, and is used for tourism purposes.

As the railroad took shape, it needed to be maintained, of course. And the area of Buford was the location for a railroad crew involved in the maintenance of the tracks and railway bed. The little community of railway workers lived in temporary housing—at first tents, later portable wooden structures—about two thousand feet from the store or business where the post office boxes were located. Thus, Buford became a railway stop, not for passengers but for workers being picked up and returned home. In between the two sets of tracks, one to handle eastbound trains and the other, westbound, was a very small depot named Buford.

The transcontinental railway workers took about two years to build about a fifty-mile stretch of railroad in that part of Wyoming. The hold-up was a trestle bridge over Dale Creek, which happens to be a creek that a person can step over. I know, because I have stepped over it! They also built the Hermosa tunnels and a pyramid as a monument to honor the Ames brothers who were the financiers of the transcontinental railroad. The pyramid is made out of granite blocks and is one of the noteworthy tourist "attractions" of I-80 through Wyoming. It stands sixty feet square at its base and is sixty feet high. It was designed by North American architect H. H. Richardson and has bas-relief portraits of the Ames brothers sculpted by Augustus Saint-Gaudens.

The Ames brothers—Oakes and Oliver—were well connected. Oakes was a U.S. Representative to the United States Congress from Massachusetts. Oliver became president of the Union Pacific Railroad from 1866 to 1869. Their family business, the Ames Company, dated back to 1774, and was well known for its steel-edged shovels. The company sold axes and shovels to miners during the California gold rush and also supplied shovels for excavating the Panama Canal, digging the New York subway, and mining Pennsylvania coalfields.

For decades, the railroad had "section crews" that lived at periodic places along the railway. The workers who lived in these homes serviced the rails and the railway bed for their designated section of the railroad. There were seven railroad-owned houses near Buford until the mid- to late 1990s. Those homes were then closed and the employees moved into nearby cities, with the railroad providing a vehicle for its employees to drive to the railroad to do their maintenance work.

An Interstate to Go Along with the Railway

The interstate known as I-80 was built in the early 1960s, and it pretty much parallels the railway through Wyoming. Before the interstate, the road was US 30—also known as the Lincoln Highway—which connected the East Coast with the West Coast. Buford was on US 30. Although many segments of US 30 did *not* become part of I-80, Buford did. In that location, US 30 simply *became* I-80. It was and is a four-lane divided highway as it passed by Buford.

When I first moved to Buford in the mid-1980s, the road down to Laramie from the summit had one lane in either direction. It now has three lanes in each direction, which allows traffic to move much faster.

In the 1980s, the terrain also was much different than it is today. The ravines between the lanes of the highway were much deeper, as were the ravines at the outer edges of the highway. Some of the drop-offs were very steep and deep—as much as fifty feet. A vehicle that slipped off the highway for whatever reason was very difficult to spot! Over the years, as the highway codes were adjusted, the deep ravines were filled in or leveled out. Filling in the ravines required that nearby hills became the source of the soil used to fill the ravines. The result is a much flatter landscape now than even twenty-five years ago.

The summit of the highway is about twelve miles from Buford, and a person can find a nice granite statue of Lincoln at that site, which is 8,247 feet above sea level.

All points east or west are "downhill" from there.

The Gangplank. The hill coming up to Buford from the east is called "The Gangplank." One of the tectonic plates of the earth slides over another plate and the result is a gentle incline. There's a turn-out and an information board explaining that phenomenon.

A few miles further to the east, about twelve miles away from Buford, is a turn-out with a sign noting that this was the location of the old Tom Horn Saloon.

The Lone Tree. The other significant landmark in the Buford area is "The Lone Tree." The tree initially was very close to the railroad and the railway workers frequently stopped to water it. The tree has been there more than a hundred and fifty years.

I have great respect for that tree. Through the years I have planted more than three hundred trees on the Buford property, and at the time I sold the store, only nineteen of them were still alive—and barely hanging on. I tried irrigation drip systems, which is what allowed the nineteen to survive. Most grasses and trees simply cannot accommodate the nearly constant winds in the area.

The wind at Buford is relentless, and often tops fifty miles an hour. In the world of trees, it is definitely a "bend or break" proposition. It is not uncommon to see trees in that area that are pretty much stripped of needles, or have very few limbs, on one side of a tree—as a result of the winds. In winter, we like to say that the snow falls horizontally at Buford. Many evergreens in the region only have needles on the "not windy" side of the tree.

The Winds and the Winter

The winds and the winter snows are a double whammy for the Buford area. In fact, if you asked me to describe Buford's environment in two words they would be "windy" and "cold."

The trees are not the only foliage to suffer from the winds.

Once you disturb the grass in the Buford region of Wyoming, it can take years for the grass to return. The grasslands of Wyoming are a very fragile ecosystem. The root system is very shallow, and it can take years for a grassy area to fill in if it has been trampled or leveled. The soil is decomposed granite, which translates into "very little top soil."

In some areas of our nation, a rancher might have as many as forty head of cattle on a few acres. In the Buford area, a rancher needs forty acres for each head of cattle!

The average wind in Buford was about thirty-five miles an hour. Many days, the winds were seventy miles or higher. I could count on the winds being more than a hundred miles an hour at least a half dozen times a year. In a word, the winds were *relentless.* It never really stopped blowing, and that is a phenomenon that can be very wearing, not only to physical structures and foliage, but to the human spirit.

More damage would have been caused by winds at that speed if the elevation had been lower. At the higher elevation, the winds inhibited outside movement, but didn't stop it. I could usually keep my feet under me, albeit my walking was a bit "drunken" in style!

A true fact: I once had a weather vane. But it couldn't survive the wind of Buford!

A Help to My Business? As odd as it may seem, the winds probably helped my business through the years. The winds come sweeping down from the summit to the west and can cause a person's gas gauge to plummet toward empty much faster than usual, especially at the high altitude. There were many times when I left Cheyenne with about a third of a tank of gas, and I'd arrive at the store with the gauge at zero.

Many of the people who got stranded on I-80 were people who simply ran out of gas. I discovered through the years that these people usually were traveling from areas in which the freeways had lots of exits, often with a gas station at every exit. That simply isn't the way I-80 works in Wyoming, but travelers who had never been there before didn't know it.

These same motorists also tended to be people who waited until they only had an eighth of a tank of gasoline left before they refilled. That also doesn't work with few exits. And it especially doesn't work if you are traveling at an elevation higher than 8,000 feet with a fierce headwind. If a vehicle routinely gets fifteen miles to the gallon, at the Buford elevation and wind speeds, that vehicle might only get six miles to the gallon.

The Brutality of Winter Winds. Winter winds were especially tough. The wind often blew at seventy miles per hour, and in the winter, the wind-chill effect could make that feel like a minus eighty degrees Fahrenheit.

Through the years, many people asked me how cold it got in Buford. I'd tell them the coldest it ever got on my thermometer was forty-four degrees BELOW zero. They found that a little hard to believe . . . but I wasn't joking!

A second main reason for people to become stranded along I-80 was this: They didn't have respect for snow. For whatever reason, they seemed to think that the snow would decrease in intensity rather than INCREASE in intensity the closer they got to the summit. They'd find themselves stuck almost before they knew they should have turned around and headed back toward Buford. These motorists tended to be people who weren't in touch with weather-warning signs or had never driven in snow.

That problem has been mostly resolved in recent years because the state has been quicker to close the road during a storm. People are told as they approach Cheyenne that the summit is closed, and they can get off the highway and find lodging there.

Unfortunately, the people who are not accustomed to driving in snow—notoriously those who were traveling up from Florida and across I-80 to get to the West Coast—were also not dressed for cold weather as they traveled the interstate. They often stopped at the store wearing short-sleeved shirts, Bermuda shorts, and shower shoes . . . in the late fall or early spring with snow on the ground!

One time I found a couple from Mississippi stranded along I-80. They were young—perhaps in their early to mid-twenties—and they were traveling with a baby in a small car. Their fuel line got frozen and there they were, stuck.

I had a big crew-cab truck at the time. I knew I couldn't leave them by the road. They would have frozen before morning. And I didn't want to spend the night sleeping in the store.

I loaded them up and told them I was going to take them to my house. They really had no other option but I feel certain they likely were very scared.

I went across the open snow-covered fields, following the ridgelines, up and over the pipeline, and all the while, it was getting dark until it was night. I tried to put myself in their shoes—there they were, a young black family riding in the middle of a snowy nowhere with a white guy, to a place they didn't know! They might have been thinking, *This guy could rob us blind, dump us and the baby out in the snow, and nobody would find us for years.* Fortunately for them, I wasn't that kind of guy!

The thought came to *my* mind very briefly that they could have done me in, but just as quickly I thought, *Naw. They'd never do that. They'd be too scared that they couldn't find their way back to civilization!*

When we arrived at the log house, I set them up to sleep in the living room. The next morning, conditions had improved and we made it back to their vehicle, dealt with their fuel line, and they went on down the road. I have a hunch they are still telling that story to their families in Mississippi.

On other occasions, there were times that I did stay at the store and allowed as many as ten people who were stranded to spend the night there.

And, from time to time, I gave lodging to people who knocked on the home that I constructed close to the store. Once, four young women were traveling and they were stranded. They weren't sure about spending the night in my house, but I think they figured that there were four of them and only one of me, and when I made it clear that I was sleeping in my bedroom behind a closed door and that they were sleeping in the living room, they made a slumber party out of it.

The winter weather played havoc, as you might imagine, with anything connected to water. The pressure pumps associated with the water well at Buford froze at times. You can't run a store with a restroom or café without water!

Having the water freeze wasn't nearly as bad as having the septic system freeze.

My home in Buford was a modular home that was well designed for the area. We were able to use *part* of the old store system for water and septic, and of course, that was an older system that could be problematic. If the store had water or septic problems, I unfortunately tended to have them at home, too.

One problem we didn't have was deep snow on our rooftops! The wind blew it away before it could pile too deep.

I noticed the longer I lived in Buford that many of the elderly people who bought in the Buford area lasted one season. That was enough "Wild West" living for them. Those who didn't like being snowbound, or staying indoors much of the time, tended to get out pretty quickly.

By the way . . .

The man who bought the town from me made statements about the wind and cold—he was reported to say that it was the coldest he had ever been. On April 5, 2012, we actually had what I considered to be a good-weather day.

I never made any weather disclosures about Buford. I wasn't asked to, and I certainly never thought to volunteer that information.

I have also noticed in recent years that the weather is a little different in Buford now, compared to thirty years ago. In the early years, it was not unusual to have four or five feet of snow out of one storm. I sometimes had to put up a ladder out at the ranch so I could start shoveling snow at the TOP of an open door, working my way down. In recent years, there were winters with less than half an inch of snow on the ground.

In the early years, the wind didn't seem nearly as strong. In recent years, there were many days of seventy-mile-an-hour sustained winds, hour after hour.

I'm not sure which is better. More snow and less wind. Or less snow and more wind.

Over all . . .

As tough as Mother Nature was at times . . .

And as wearing as the winds and cold might have been over the decades . . .

I never once thought about moving back to sunny southern California. I was determined to beat Mother Nature, and once I came to the realization that I wasn't going to *win*, I was determined to co-exist, compromise, or call it a draw.

5

ROOTS

THE REASON TO BUY A TRADING POST

There's always a *reason* for everything, whether we know it or not. The main reason I ended up being a *storekeeper* in Buford was my son Jonathan.

I neither blame him for that fact, or credit it to his account. The truth, however, is that I needed to put down roots and the reason for the roots was Jonathan.

Jonathan was born in April 1984, just about five months before we moved to the ranch and our log home there.

The story that led to the purchase of the trading post is an intensely personal one. It is also part of the story that led to the eventual sale of the store.

A Matter of Custody

I knew that it would take a unique woman to live at Buford. Jonathan's mother had been that type of woman. She loved the outdoor life, loved the mountains, loved having a garden, and loved riding horses. She could go out and shoot a rabbit, skin it, and put it in a pot with some freshly harvested vegetables from her garden, and be very pleased both with herself and with her homemade stew! She did not need lots of shopping, outings with girlfriends, or elaborate appliances or luxuries.

Unfortunately, she did need alcohol. That need had turned into an addiction long before we moved to Wyoming, and very likely, was an addiction long before we first met.

As time went on, Terry struggled to stay sober, and especially after she made the decision to stay in Buford and not go out on the road with me.

In the beginning days of living out at the ranch, I was still doing a lot of hauls for Allied that took me miles away to new places. I often was gone a month at a time. I custom designed my truck so that it had a shower and a stove. The idea was that Terry and the baby could come with me. I had a child seat built into the cab so we could travel as a family.

We actually did that for a couple of years, and then she made a decision to stay home. I think part of the reasoning behind that was a desire to drink. She knew that it was not EVER going to be possible for her to drink if we were out on the highway. I didn't drink. I didn't stop at places where she could drink. She *had* to be sober if she was going to sit with Jonathan and me in the cab of my moving van.

Having the Sno-Cat by that time made it easier for her to stay home. She had accessibility to Cheyenne and the substances she could find there.

The more she stayed at home, the more I came home to find her drinking. I'd explain to her that I couldn't leave her at home if she was going to drink because I couldn't count on her being sober as a mother to our son. She would agree to go back into rehab, and while she was in rehab, I'd stay home to take care of Jonathan.

Certainly, I knew that Terry had a drinking and drug problem before we moved to Wyoming. She struggled with it. Like many people, I think we both thought that a change of location and a "fresh start" would make a big difference. We hadn't yet learned the time-proven principle: "Wherever you go, there you are." Circumstances and geography don't dictate addiction. Addicts can always find a way to fuel their addiction.

When Terry and I married, she had been clean and sober for several months. Drugs had been easy for her to acquire in St. Louis and in the Los Angeles area. She knew that I wasn't at all into drugs and she, too, wanted to be free of them. During our marriage, however, she traded her drug addiction for an alcohol addiction.

An addiction of any kind is a terrible bondage. Not only for the person who is addicted, but also for that person's entire family. In the early days of our dealing with Terry's addictions, we didn't know that alcohol addiction is a "disease." From my perspective, it was a habit that could be curtailed by the human will. A tough habit, to be sure. But a habit. It was only in later years that I came to recognize that Terry had a serious disease that required outside help. She likely had inherited that disease, or at least a propensity for it, from her mother, who was also an alcoholic.

In all, I put Terry through treatment five times after we were together. The sixth time she showed up at the rehab hospital, they refused to admit her. They said, "You are wasting your time and money, and our bed."

Shortly after Jonathan turned five, Terry left and took Jonathan with her.

I was convinced then, and remain convinced now, that Terry didn't leave Buford primarily because of Buford—the isolated ranch, the weather, and so forth. Neither did she leave *primarily* because of me. She left, I believe, because she was tired of trying so hard and for so long to be sober, and with so little support. She had relatively few friends in Wyoming, and on a trip to St. Louis, she rekindled relationships with a few high-school friends who still lived there—one man in particular. She went back to St. Louis to be with him, and she took our son with her.

I didn't want to be married to a person who wanted to be with someone else, or to live a lifestyle related to substance abuse. I agreed to the divorce, but not to Terry having sole custody of Jonathan.

During the divorce hearings, my attorney suggested that, given Terry's history of alcohol and drug use, I might want to have her tested randomly. It was an issue that could be used to gain full custody. I didn't think Terry was using at that point, and when she readily agreed to have a random drug test, I made the assumption that she was clean and sober.

I also knew that the drug testing would stall the divorce proceedings, and be an extra expense. We were both eager at that point to get the divorce behind us. So I declined to have that done. It was a big mistake on my part.

The truth started coming out a couple years after she left. The daughter-in-law of the people who owned the Lone Tree

Junction Store had two children who were about Jonathan's age. Her children played with Jonathan and she became a friend to Terry. She perhaps was the only person who really knew how much Terry was drinking, and how many men she had found for "entertainment" in Cheyenne while I was on the road. But . . . she didn't tell me until long after the divorce.

In the end, I was awarded *shared* custody of Jonathan in the divorce, but I learned very quickly that it was difficult to establish that kind of custody across state boundaries. There was virtually no "sharing."

Jonathan's Early Years in St. Louis

By the time Terry left and moved back to St. Louis, her brother and sister seemed to accept the fact that alcoholism was ingrained as a part of Terry's identity. There were never any attempts at intervention that I knew about. I suspect that Jonathan concluded matter-of-factly that "drinking" was just something that his mother did, perhaps even assuming that all adults drank as she did.

I knew, of course, that life in St. Louis wasn't all roses for Jonathan. I'd get word from time to time that his teachers were concerned that he was coming to school day after day in the same dirty clothes. He was "acting out" in inappropriate ways at school and on the playground.

Over a five-year period, I began to pursue FULL custody in the court system of Wyoming.

Did I try to have a relationship with Jonathan during the five years he was away? Absolutely.

I called a number of times, only to be told that Jonathan wasn't at home. I tried to arrange for visitation—either my going to St. Louis or bringing Jonathan to Buford for a week or two of vacation—and each time, my efforts were negated.

I went to court three times in the five years he was away to try to gain full custody, and at the bare minimum, regular visitation privileges. Each time, the Wyoming court would make a ruling but the Missouri court refused to accept any decisions made in Wyoming. The Missouri courts demanded that Wyoming courts step aside, and of course, the Wyoming court officials refused to do that. It was a state-to-state stalemate and I was hung up in the balances.

The divorce decree stipulated that any time Terry was incapacitated or ill, Jonathan was to be sent to me. The problem with that? Nobody ever admitted that Terry was incapacitated or ill! On one occasion she was hospitalized for treatment of hepatitis and Jonathan spent his days with her boyfriend. On other occasions, other family members stepped in to provide care. It was only after I received medical bills from a hospitalization—owing to the fact that I was providing health insurance for Jonathan on a family plan that inadvertently covered his mother—that I learned that she had been hospitalized. All after the fact, of course.

On one occasion after a Wyoming court ruling, I went to pick up Jonathan. I called the police in advance so they could go to the house with me. They took a look at the paperwork I had brought with me and announced that it didn't apply to Missouri, so I had to return to Wyoming without my being given permission to even see Jonathan.

On another occasion, I had letters from Terry's mother and several other people—including the principal of Jonathan's school—all in support of my having custody of Jonathan. But when the decision finally came through, most of these statements were ordered to be kept out of the court record. Terry was allowed to make her statement over the phone, but all statements on my behalf were denied as evidence.

Was I frustrated by all this? You have no idea.

As I learned more and more about Terry's lifestyle in St. Louis—mostly about her resumed drug use—I sought legal counsel. My attorney advised me that it would be virtually impossible for me to gain actual physical custody and to bring Jonathan back to Buford as long as I was away from Buford much of the time in the long-haul moving business. I had a good income and a long track record as a stable parent and provider, but I was also "absent" for many days, sometimes weeks, in any given month.

My attorney advised me to get a job, or establish a permanent business located close to the ranch.

About the same time as I was seeking to bring Jonathan back to Buford, the family that had owned and operated the Lone Tree Junction Store had decided to sell it. The husband and father of the family had died and his widow wasn't interested in trying to run the store on her own. The children had no interest in running it. And very quickly, we entered into negotiations for my purchase of the store. I had lots of ideas about things I might do to make it more profitable, including some marketing ideas for attracting more customers. The ownership of the store was a challenge!

I knew about general business matters from the moving operation that I had been running for more than a decade. I didn't know anything, however, about running a gas and diesel outlet, managing a postal service, having a café, or running a retail business.

Nevertheless, it was ROOTS. And I needed ROOTS if I was to have custody.

Jonathan's Return to Buford

By 1995, I still did not have custody of Jonathan. That year, Terry and Kevin got into some kind of domestic argument that resulted in the police being called. Child protective services entered the picture in St. Louis. Nobody in St. Louis had ever told these government caregivers about me!

Prior to that time, when the child-protection-services personnel had been called in—on a number of occasions—they would always call Terry, announce precisely when they were coming out for a home visit, and Terry would immediately clean up the house, get her ducks in a row, get sober, and be ready for their visit as an exemplary model of motherhood. It wasn't until things turned violent and the police got involved that the child-services officials in Missouri finally agreed to hear the requests of the officials in Wyoming and Jonathan was sent to live with his father in Buford.

When the child protective services in St. Louis advised the courts that Jonathan should be sent to Wyoming to live with me, he didn't want to come. I could understand this because he had been away from me for nearly the entire six years since Terry had moved from Buford. He had been told all kinds of lies about me, and the last thing he wanted was to be sent off to the middle of nowhere with a man he didn't really remember!

Starting "New" in Buford

When Terry left with Jonathan, he had nearly finished his kindergarten year at the little school first established for the railroad-worker children in Buford.

A building I called the "Old Schoolhouse" was located on my property. I used it as a personal storage and office area. The schoolhouse had been established primarily for the children of the railroad workers and was in operation from 1905 to 1966. A new schoolhouse was built a few miles away when the population and needs of the school outgrew that original building. That was the schoolhouse of Jonathan's childhood.

Jonathan came back to live with me at age eleven. By then, he was in sixth grade. And by then, the school was located eleven miles away. It was a two-room brick school with a teacher, who also served as principal, and an assistant. One room was used for science and history, the other for English and math. The building had a basement that served variously as a little auditorium, an indoor exercise area, and a lunchroom. Even though there were only a dozen or so students in attendance, the facility had all the basics of a much larger school. The bookmobile came regularly for access to library and research books. Music and art instruction were provided by visiting mentors.

For junior high—from seventh grade on—and high school, the students went by bus to Cheyenne.

I thought the education offered at the little school was excellent. There was lots of individual attention, the camaraderie among the students of various ages was an excellent mirror of life, and the emphasis was on personal pacing and lots of reading and research.

My hope was that Jonathan would find a sense of peace at the little school and his reestablished life in Buford. His time in St. Louis had been chaotic. The school there was large and his life had a lot of instability to it.

Unfortunately . . .

Jonathan responded to much of his life in Buford with a fairly intense variety of pre-teen and then teenage rebellion.

Our school-day routine involved my fixing breakfast for Jonathan and getting him on the bus for a nearly hour-long ride to the junior high school in Cheyenne. That long ride was reversed at the day's end, which meant that he returned home in the late afternoon. In the winter months, the sun might already be setting. The store was usually open for another hour or two and Jonathan would be there with me. And then, we'd go home for dinner, homework, and bedtime. It was a highly regimented schedule.

A fixed schedule was not at all something to which Jonathan was accustomed.

He also was not accustomed to being given any chores or responsibilities, no matter how minor. He wasn't used to a father who believed in the value of chores.

I gave Jonathan simple tasks, such as sweeping the floor

or taking the trash out at home or in the store. I'd nearly always find that the trash had been thrown down the ravine behind the house rather than taken to the dumpster a few yards away. Or, I'd find a couple bags of trash behind his dresser. The floor in the store would have things swept under the tables, not swept into dustpans and put into trash bags. The chores, of course, were not considered punishment of any kind on my part—they were chores intended to instill a sense of discipline and responsibility in a boy on the brink of becoming a teenager.

Clothing became an act of rebellion for him. Jonathan didn't like the clothes I chose for him. Jeans. Nice T-shirt or regular shirt. Nothing out of the ordinary from the other students. He developed a habit of taking some of his St. Louis clothes with him in his backpack, and changing clothes between the time I put him on the bus and the time the bus arrived at school.

Jonathan never ran away from home but he certainly was living in the past.

And his past had been a life of living with alcoholics and drug abusers—both Terry and her friend Kevin.

Jonathan was only displaying behavior problems in Buford that he had previously displayed in St. Louis.

When Jonathan came to Buford, I had several conversations with him about his opportunity to have a "fresh start" at the school in Buford. "Nobody knows you here," I said to him. "You can be whoever you choose to be." I reminded him of some of his difficulties in St. Louis and said, "Now is the time you can make a positive change."

Unfortunately, Jonathan didn't have much experience in

what might *create* a positive impression or good relationships with his peers. I heard one time about a woman who told her son to "be good" several times as they prepared to attend a party. The boy finally said, "I know you want me to be good but *how* do I be *good?*" Perhaps if I had sat through several days at the school with Jonathan and watched how he interacted with the teachers and students, I might have been able to give him some practical advice, but that's in hindsight.

The biggest strike against his success back in Buford was that Jonathan really didn't want to be there. He had been told a number of lies about me, over a period of more than five years, and those lies *were* his perception of me and they lay at the foundation of our father-son relationship. Jonathan not only found himself at age eleven in an environment he barely remembered and that was foreign to the previous five years of his life, but he was in that environment with a father he barely knew and didn't trust.

It was a difficult time. And that's a huge understatement.

A Move to a State Institution

Even after Jonathan came to live with me, I was under close scrutiny by child-protective services in Wyoming. The main reason was that Jonathan apparently made statements to these officials that he hoped would cause the officials to allow him to return to St. Louis. The officials never accused me of child abuse, but I suspect the officials drew that meaning from what Jonathan said to them. No proof. And I hasten to add, I never abused my son. In the end, I was never declared to be an unfit father.

From Jonathan's perspective, his *living in Buford* was a form of abuse. He didn't want to be there and he was determined to do whatever he had to do to be sent away.

Jonathan was a smart kid, and he decided that if he could make me look like a bad father, the court in Wyoming would send him back to his mother, which is where he wanted to be. Terry may not have been the most stable woman or the best mother, but she was the only mother Jonathan knew and she, in essence, *was* his family reality.

What Jonathan hadn't counted on was that there was a third option—if he wasn't with me, he wouldn't necessarily be sent to Terry. The court might order him to a facility for troubled children right there in Wyoming.

And that's what happened.

Child-protective service officials in St. Louis had sent Jonathan to Buford. State officials in Wyoming weren't going to send him back. They ordered him to a state institution at Casper, Wyoming.

In Casper, the state officials insisted that Jonathan be kept as a full-time resident. I frankly didn't question the decision. I knew my son needed emotional help and I trusted the officials at this institution to be professionals who could give that help. If they stated he needed full-time residential care, who was I to argue?

In retrospect, I perhaps should have argued. Then again, in reflecting on that decision, there would have been the potential that Jonathan might have run away, solicit a ride from someone on I-80, and end up hundreds of miles away under the control of a total stranger.

The state informed me that I could make a choice: either pay for Jonathan's health insurance, which would be billed by the state for psychological and physical health-care services, or I could drop my family-plan insurance policy and pay "child support" directly to the state. I opted to keep the insurance.

It was more than irony, I believe, that the week the insurance money ran out for Jonathan's institutionalized care, the state institution announced that he was ready to be released. Within five days of my telling the state officials that the insurance money was fully depleted, Jonathan was transferred fully to my care. (The state had full custody and I had temporary custody—and that was reversed in a day.)

When Jonathan was initially institutionalized, I asked how long he would be there. They told me four to six weeks.

He was there for two-and-a-half years.

During his stay there, I drove up to Casper—almost three hours away—every weekend. On some weekends, I'd get a hotel room and we'd spend a couple of days together. He usually spent Saturday night with me in the hotel. That gave us time to do various activities together, share meals, and talk. I was not allowed to meet with his therapists or to participate in joint family counseling. I found it rather weird that these officials would agree to my taking him "off campus" but not participate in his therapeutic sessions or have any form of custody the other five days of the week.

A few times we went back to Buford if it was a long weekend or a holiday.

The long commute to Casper was worth it to me so I could spend time with my son and try to establish something of a normal routine and normal relationship with him.

The Rebellion Continued. The state institution where Jonathan was kept was a lockdown facility. He was the only person ever housed there who broke out of the facility. He managed to jam the lock of a door and let himself out late at night. He was picked up later after hours of wandering around Casper. I was called by those in charge of the institution to see if he was with me—otherwise, I doubt that I would ever have been informed that he had outsmarted their protocols for security.

Behavior such as that, of course, convinced the state officials that he was still in need of their psychiatric services!

Resentment Compounded. Jonathan was in state custody when Terry died. I was infuriated that the state officials told him about her death rather than letting me talk with him.

A girlfriend of Terry's heard of her death and called the state hospital and told them what had happened. They, in turn, told Jonathan. He called me in tears to find out if it was true. I made calls that revealed the information was correct, and I also learned how he had been informed.

None of this spelled good parenting, or appropriate medical protocols, to *me*!

As I stated earlier, when Terry moved out, before the divorce, she returned to St. Louis to live with an old boyfriend named Kevin. He had been a drinking and drug-use buddy in high school. She came home from the reunion with her mind made up that she was leaving me and going to live in our old hometown.

She certainly didn't want me to come along. She loaded a U-Haul truck with some of her possessions and off she went.

She stayed with Kevin and died with him.

One night Kevin and Terry sat on her sofa and drank shot after shot of hard liquor until they both passed out from alcohol poisoning. Kevin died that night. Terry was taken to a hospital in a coma, and died seven days later.

It was a sad end.

Terry's death was a major blow to Jonathan. He believed that if he had been there with her in St. Louis, she would not have died. And in the end, that meant that he felt a false form of guilt over her death.

It also meant, of course, that any strides we had made toward a healthier father-son relationship came crashing down.

The Anger Continued. Once Jonathan was released and came "home" to Buford, his anger continued.

In the towing business, I had acquired a couple of cars that were actually in fairly good shape and required very little to repair. Some of the cars I towed were abandoned, and the rules allowed for me to acquire the title to those cars if nobody called to claim them.

When Jonathan returned to Buford, he was sixteen. He had been away nearly four years. I taught him how to drive, and I let him choose a vehicle from the salvage yard I had behind the store.

When I was a teenager, I resented my father choosing a car for me without any input from me. So, I pointed out to Jonathan a couple of cars in the salvage area that we might fix up for him to drive—but I clearly left the decision to him about which car that would be. In the end, I'm not sure my benevolent approach mattered. Jonathan didn't seem to want *any* vehicle—or pretty much any other form of *gift*—that came from me. He did, however, choose a vehicle that he might drive in the course of *learning* how to drive and practicing the necessary parking, steering, and other skills important for getting a license.

The vehicles became a way for him to express his anger—violently, I might add. I recognized head-on that he was still struggling with major emotional issues.

There would be times when I'd see him "practice" his driving on the Buford store property—out behind the store and our home. I watched from inside the store. He'd be doing just fine and then suddenly, it was as if something came over him and he went berserk, backing into trash cans, hitting other things on purpose. The end result was a "crashed" car, not just an abandoned vehicle. On one occasion, he did significant damage to the roof of a car in the salvage area, including breaking out the windshields.

I was at my wit's end, and after getting counsel from several child "experts," I made arrangements for Jonathan to be placed in a home run by the Catholic Church. It was well known for helping troubled teens and it was significantly closer to Buford. He lived there for another year and a half.

Letting Him Leave at Age Seventeen

Jonathan was seventeen when he returned home from the Catholic institution. He wanted to drop out of school, leave the area, and move to Portland where he could live "on his own." I wanted him to stay until he was eighteen. The more counselors I consulted, the more I heard them say, "What's the point?" A few months weren't going to make any difference in the outcome, and forcing him to stay would only have created deeper ravines of resentment.

I got the opinion of the police, in part to determine my legal rights and limitations. They told me that, in their opinion, I should let him go. They told me that they could follow him, arrest him, and bring him back—but that he was likely to leave again. They suggested that I would be better off letting him go, insisting that he pay his own way, and also letting him know that I wanted to stay in touch and would always be concerned about him.

Frankly, who was I to talk about staying at home? I had left when I was *sixteen*, and my parents had wisely let me go.

I also tried to put myself in Jonathan's shoes and when I did, I realized that I would want to leave, too! All of his friends were miles away, and he was pretty much stuck in the middle of nowhere as a seventeen-year-old. He had no interest in running the Buford Trading Post, which I had secretly hoped might become his ambition.

He was an artist and was attracted to the artists' community in Portland. And indeed, that is where he went. He exercised a significant amount of ingenuity there in getting a job and place to stay, and supporting himself.

He lived in Portland for almost three years, perfecting his artistic skills, especially his pottery skills. In my opinion, he's an excellent artist, but unfortunately, there's not a huge demand for potters.

I made an offer to Jonathan when he was in his mid-twenties. He took me up on the offer and enrolled in a junior college program. I gave him a set stipend to cover the basic costs of living and going to school to finish a degree.

He now lives and studies at a place called Sunrise Ranch in Loveland, Colorado. It is a communal living community with a strong vocational program. The ranch curriculum includes farming and mechanical skills, computer skills, art training, and other practical skills that prepare the students to work either at the ranch (as instructors or workers of ranch businesses) or in other cities. The ranch has a good reputation, has a religious component to it but is not considered a cult, and has a disciplined lifestyle.

We are much closer than when he was a teenager. It took a while for him to sort out lies from truth, and to see the value of discipline, chores, and responsibilities. We try to see each other once a week and talk fairly often on the phone. In many ways, I feel as if we are *finally* forging the father-son relationship I had always hoped for. We're having opportunities to explore some of the old hurts and disappointments, and to make sense of them.

And why am I telling you all this?

Part of the reason is to let you know that there are no "ideal environments" that give immunity against family troubles, addictions, divorce, or emotional illness. Problems exist in all shapes and sizes.

The image of the West as a place of total self-determination, rugged wholesomeness, and family values is strong. Those traits don't always play out in the same way for everybody.

Addictions can be devastating to a family anywhere, including the Wild West.

Children can go through very traumatic and troubled times anywhere.

And in the end, love is powerful . . . anywhere.

6

THE STORE

BUYING IT, RUNNING IT, LEARNING FROM IT

The official name of the store closest to my ranch on I-80 was the Lone Tree Junction, but that name was always a mystery to me. The "Lone Tree" monument—commemorating the hardy long-lived tree that had been a landmark in the region—was about three miles up the road. At the exact site of the Lone Tree Junction Store there was neither a tree nor a junction. The nearest highway intersection was also miles away.

Shortly after I moved to the ranch, I realized that nobody called the store by its official name. Everybody just referred to the place—and thus, the store—as "Buford."

After I purchased the store in 1992, I kept the Lone Tree Junction name for about a year. Every time people would ask me, however, where I lived or worked, they'd look puzzled when I said "Lone Tree." When I started saying "Buford," they'd reply, "Oh sure, I know where that is."

Regardless of the name, I knew the Lone Tree owners made money at their store. Their main customers were the railway workers who lived nearby. They bought their fuel there, and in the evenings, the railway workers often came to the store to drink beer and shoot pool.

I was one of their customers, buying fuel from them and a variety of sundries.

I also knew that they didn't really run the store *as a business.* They ran it as a convenience store for those who *needed* to stop for fuel or supplies. The husband had worked as a mechanic for a major airline and when he retired, the family had moved from New Jersey to Buford. He loved the outdoors and wanted to hunt and fish with his son in his retirement years.

The family lived on site, and in many ways, they had a home-based "store" to help out their neighbors.

As a customer I'd often find myself questioning, *Why don't they do THIS? Why haven't they tried THAT?*

I think it was my own sense of "what if" that made me interested in taking on the challenge of the store. It seemed to me that there was greater success to be made of this isolated quick-stop.

I didn't fully recognize at the time that the previous owners were living the vision they had for the store. They wanted a lifestyle that allowed them to hunt and fish when they wanted to. They had a more relaxed approach, with perhaps less pressure or self-ambition to build a highly profitable business.

I know the husband enjoyed running the Lone Tree Junction Store. But I'm not sure his wife did, or that their children did. Their son could have taken over management of the store—he had lived there with his parents for ten years before his father died. The son opted to have a job where he could work nine to five. For a couple of years, we invited the family to our house for Christmas Day. To my surprise, they came—but in shifts. There was always somebody back at the store, because they had a fear that if nobody was there, the store would be burglarized. (They had a living room and bedroom at the back of the store, and used the kitchen that served as a store café as their own kitchen.)

I had no interest whatsoever in living with that fear. I decided even before I purchased the store that if I ever owned the store, I'd put in an alarm system and have sufficient insurance to cover any loss, thus giving myself permission to leave the premises.

It was that family fear and decision to "always have somebody there" that kept them bound to the store. A little more "away time" might have meant that the son would have been more interested in taking on the business.

But to each his own. In the end, my son didn't want to take over ownership of the Buford Trading Post, even though I had hoped for that.

My vision was aimed at maximizing the full potential of the store *as a business*. I wasn't retired and didn't want to be retired. I desired a challenge and an adventure—and perhaps it all came down to, "you get what you are aiming at." The previous owners and I aimed at two different goals, and in our own ways, we both succeeded.

Moving into a New Business!

The Lone Tree Store had been built in 1939. It was a good building, but it was an *old* building, and it had all the problems you might imagine associated with age. It rested on 9.9 acres. A building affectionately dubbed "The Old Schoolhouse," built in

1905, was at the back of the property. When I purchased the property, I turned it into an office and storage area.

I finally reached an agreement with the widow who had inherited the Lone Tree Store and I made arrangements for the $150,000 sale price and got the keys to the buildings.

I took possession of the Lone Tree Store just a day or two before Christmas in 1992. I spent my time scrubbing floors and washing the grime from windows that probably hadn't been washed in twenty years. The buildup of dirt and cigarette smoke was amazing.

I was determined that when the railway workers came into the store after Christmas, they'd see a change. I wanted them to walk in the door and think, *Something has changed!* I didn't need to advertise "under new management."

The Challenge of "Sole Ownership"

When you are the sole owner of anything, you tend to wear a lot of hats. Based upon my experience, the sole owner of a "trading post" store tends to wear all those hats simultaneously.

On any given day I was a short-order cook—serving breakfast and lunch at a small counter.

I was the postmaster—at first, sorting mail into pigeonhole boxes and selling stamps for an hour each morning, and later, up until noon.

I was the stock boy and shelf-keeper for items that would typically be found at grocery stores, hardware stores, and drug stores.

I was the cashier for the store and the gas pumps.

I sold souvenirs and regaled customers with tales of Buford in years past.

I gave advice, as best I could, to those wanting to know about the sights in the area, the weather expected during the next two hours, and the nature of the "car problem" that sounded like ka-thunk-ka kunk-ka-thunk.

I mopped the floors, washed the windows, kept the restrooms clean, and after hours, placed orders for supplies, kept business accounts current, and occasionally when an employee was present and there was a lull in the day, made a run to Cheyenne for items necessary for the ongoing operations of a roadside business.

I also took a more aggressive approach to "marketing" and advertising than my predecessors and some others who ran businesses along I-80.

It's one thing to have the only store along a fifty-mile stretch of highway. It's another thing to have people stop at that store and actually *buy something* while there.

I had a lot of learning to do.

And a lot of decisions to make. From the first month, I made decisions about some things to keep and some things to give up.

Food Service. Food services had been available in Buford but the café operation had been pretty much limited to the owners setting up a plywood plank to set up a buffet for the railway workers.

I put in tables and chairs to seat twenty-eight, and I offered food service from 7 AM to 3 PM. I served breakfast and lunch, and by closing at three o'clock I had plenty of time to clean up and get ready for the next day. Initially, I hired a cook to make lunch specials.

The menu was simple and hearty—perhaps spaghetti and meatballs one day, a half chicken and mashed potatoes, meatloaf and potatoes.

We also had basic breakfast foods, and after the lunch hour, we could make hamburgers and French fries for customers who wanted them.

The kitchen and eating area was separate from the rest of the store, which made it much easier to deal with cleanup and food-service hours.

During the summer months—basically from Memorial Day Weekend to Labor Day Weekend—I hired a cook from the university to work for me. (The university cafeteria was closed in the summer.) When he returned to the university in the early fall, I made some menu revisions and offered only things that I could prepare—burgers and fries and hot dogs and so forth. That worked well for a number of years.

Bar Service. Even though the "bar" at the store had been primarily for railway workers who lived just a few dozen yards from the store, I also made a decision to shut down the "bar." I didn't want anybody stopping for alcohol and then getting back out on I-80. I sold beer out of the cooler and by the case, but not as an on-tap beverage.

Operating Hours. When I first began operating the store, I was open from seven in the morning until eleven at night.

Later, I also began closing the store at eight o'clock in the summer, and at six o'clock in the winter.

The only time I closed the store was on Christmas Day, and on days when the highway was closed by the state owing to weather. I was open on Christmas for the first few years but realized that I didn't have enough business on Christmas to cover the cost of the lights. In the later years, I also closed on Thanksgiving and New Year's, but that was after I had a pay-at-the-pump option, which allowed travelers to get the fuel they needed even if the store was closed.

Also in the later years, the restrooms were open twenty-four hours a day with an external entrance, and there were vending machines outside for soft drinks and a few snack items.

Fuel. The previous owners left me a short list of vendors, including the name of their fuel supplier. I made a decision NOT to use their fuel supplier and to go with someone else.

A Big Increase in Inventory. When I took over the store, I had two coolers for soft drinks and beer, and a candy corner, and a small display case with a few trinkets in it.

I felt so busy those first few weeks just keeping up with that level of inventory!

As I look back, I realized I was much more relaxed and seemed to have much more time by the time I was ready to sell than at the time I first purchased the store.

I learned how to become increasingly efficient over the years, juggling many more streams of income and a much greater inventory.

Always Something to Do or to Fix

I think I would probably be a lousy employee. I would be a "good worker," and I would have high job attendance, but I likely would be told numerous times, "You aren't following directions!" I would find it difficult not to try to put my own twist on things, especially if I saw a way that I thought something could be done better. I likely would always be adding my own vision.

If I wasn't an entrepreneur before Buford, I am now. I can't imagine being anything other than self-employed.

Some of the people who came into the store really seemed to envy my life. I heard from several people after word got out that I was selling the store. They wrote, "You *can't* quit. You're living the ideal life."

I discovered that these people tended to envy my life because they saw me as being very "free"—free to do what I wanted, when I wanted, and to the degree I wanted. Actually, that was far from the reality.

Customers who regularly came into the store to get coffee and pay for gasoline at eight o'clock in the morning on their way to work or school . . . expected me to be open at eight o'clock!

If I got up in the morning feeling a little subpar, I still had to get over to the store and run it. I never had the privilege of "taking a sick day" and spending the day in bed, and still getting paid for it.

If I wanted to go on an extended vacation, which was more than a few days, I had to plan for it—not only set the dates, but arrange to have someone on site managing the store, have sufficient product ordered in advance, leave checks to be mailed at certain dates, and so forth.

There were certain things that needed to be done every day at the store, whether I wanted to do them or not.

For example: After I rebuilt the store, I had the restroom entrances put on the outside of the building, and I purposefully left them open around the clock. I wasn't at all interested in having someone knock on the door of my house solely so I could let them into the store's restrooms!

For whatever reason, some people have no sense in what they attempt to flush down a toilet. Every morning, one of the first things I did was go to the restrooms and flush the toilets, and make sure that nothing was backed up.

If there was a problem, that toilet became a top priority in my day. There was no way around it.

I learned how to run an auger for myself so I didn't have to call a plumber.

It was always on the hottest summer day that the ice machine seemed to break. And especially so, it seemed, if the nearby campground was filled to capacity. Well, I learned how to fix the ice machine. Eventually, I bought a new ice machine and installed it myself.

And, if I couldn't fix the icemaker for some reason, I'd call the local ice company and ask them to bring me a hundred bags of ice. I wasn't going to disappoint those customers, or miss out on that income.

If I had no overnight gasoline sales, I had to suspect that the pumps were broken. This happened sometimes in the wake of an electric storm. The pumps were designed to shut off in a power surge, and then reboot. But at times they failed to reboot. They were waiting for a manual punch of the "reset" button.

I discovered that when the wind would blow really hard for several days in subzero wind-chill temperatures, I tended to have difficulty with my furnace. I finally diagnosed the problem. The wind blowing over my chimney would cause the snow that had been melted by the heat from my fireplace to drip back down into the air intake vents on the roof, and freeze them shut. That meant that there was no fresh air to feed the furnace, and when that happened, the furnace would shut off.

The net result was that I would awaken early in the morning to find no heat in my home. And the only solution was to get the ladder, climb up onto that icy roof, and chip away the ice. There obviously were no manufacturers who had factored in the weather conditions of Buford! I eventually rebuilt the vent system so the water didn't drip into the vents . . . and solved the problem.

One year while I was on a trip to Alaska, the gal I had hired to manage the store in my absence called me to ask a question about the icemaker. I walked her through the repair she needed to make. She also informed me during that call that the man I had hired to be her assistant had quit just hours after I left.

I offered to return home immediately but she encouraged me to continue with the trip and told me that she'd hold the fort until I returned. She did a great job of it! She worked long hours and I paid her twice what we had agreed upon—she had done the work of two people.

I don't think—and never did think—that there is a "right" or "wrong" way to run a small business like the Buford Trading

Post. Nobody is likely to be an expert in running this kind of business, at least not in the first year or so.

What did exist was a "Don" way of running the Buford Trading Post. It would be a different way for another person.

Many of the lessons I learned from the store, and AT the store, came down to one word: RETAIL.

7

RETAIL

YOU BUY STUFF. YOU SELL STUFF.

For a while, just prior to completion of my purchase of the Trading Post, I got cold feet and was on the verge of backing out of the deal.

I admitted this to a friend, who happened to be the Allied agent to whom I was "assigned." I said, "I don't know what I've been thinking—I don't know anything about retail!"

My friend replied, "Aw come on, Don. What's so hard about this? You buy stuff. You sell stuff."

And for whatever insane reason, that was enough logic to get me through . . . well, that and the fact that my attorney advised me that if I backed out, I was likely to be sued by the seller.

And so it was—I plunged into the business of running a retail store!

I was keenly aware that I had to *learn* the retail business. I didn't really have any expert role models and I had no experience in retail. I discovered that there were numerous trade magazines related to retail and I relied heavily on them to learn what millions of dollars of marketing research had revealed. The magazines were filled with articles about what worked and what didn't work.

Unfortunately, for a variety of reasons, what worked for many stores simply didn't work in Buford. Many times that was owing to location, which directly impacted the freight costs associated with the delivery of goods. Sometimes there were other factors I could isolate. I decided I needed to become my own "marketing research" team of ONE!

I put myself on a short leash. I gave myself thirty days to determine if a product was going to be successful in the store—in other words, I only purchased a limited amount of *anything* at the outset, and after thirty days, if the product seemed to be moving, I would reorder. If not, I'd do my best to get rid of what I had, with no reorders. This meant, of course, that I was nearly always in the process of reevaluating *something* on my shelves.

I had my accountant give me a P&L every month, rather than every quarter as most retailers. He itemized about a dozen items so I could tell quickly from month to month if my sales in each category were increasing or decreasing—and then, if decreasing, attempt to determine WHY.

Part of the reason for the thirty-day window was my short high-selling season—from Memorial Day to Labor Day. I simply couldn't afford to wait three months to evaluate the saleability of a product. For me, three months meant a rollover to the next summer.

I also was limited, of course, by the size of the store. There's only so much variety that you can cram into a couple thousand square feet.

It was a massive exercise in trial and error.

And most of the time, all this was fun for me. It was a

game and an adventure—a challenge I enjoyed. It kept my mind actively engaged in the *business.* I called it the "sport of retail."

Challenge #1: Get Them Off I-80

I faced several challenges when it came to getting customers and making sales.

The first challenge was to get the potential customer off I-80.

Gas was what brought most people OFF I-80, so that became the first-tier challenge for me. You can't make a sale if the customer doesn't make a stop!

Long-distance travelers were very good gasoline and diesel customers, of course. Especially in the winter months, I could nearly always count on "fill up" from a customer pulling in for fuel.

The gas price for long-distance travelers was nearly always the full price. I made an exception for skiers in the winter months. I'd tell them that if they stopped back at the store after skiing, and they brought with them a lift ticket, I'd sell the gas at a discount of a few pennies per gallon—which was the price of gas at most outlets in Cheyenne. The point was to get them to stop on the way back, hoping they would buy something in addition to gas. It didn't matter, of course, *which* lift ticket. I wasn't locked into Steamboat Springs.

A study came out that about 13,000 people were commuting from Laramie to Cheyenne every day down I-80. I made a conscious decision that I wanted the "local commuters"!

Most of those commuters were going to and from work, or

from the college. There had to be a *reason* for people to stop midway between these two cities to get sundries or purchase fuel, rather than wait another half hour to do their shopping in one of the bigger cities.

I wanted some of those people as regulars! I was painfully aware that gas prices were lower at the larger fuel outlets in Cheyenne—usually a few pennies per gallon lower. Why would anybody want to stop at Buford and pay a little more? Well . . . I made it my mission to give them a reason!

First, I decided to match the in-town price for commuters. My gas pumps would not allow for me to automatically calculate the difference in price just for commuters. I had to figure that by hand. It was laborious, so I sought another way.

Second, I developed a system of giving to returning customers a little ID card that had a customer number on it. These were given to local residents to encourage them to come to ME for fuel. Each time they stopped to fill up, I'd write down the number on their card and put it in a fishbowl, and at the end of each month, I'd have a drawing to give away something of value.

I approached various vendors and they nearly always would give me at least one prize a year. The prizes were gender specific, however. Half of my customers, of course, weren't always interested in receiving the prizes I gave.

I finally settled on giving a prize of a U.S. Savings Bond.

I eventually distributed about two thousand cards with ID numbers and, in a way, it was good advertising even if the people who had those cards didn't always use them.

I also developed a "spinning game" for commuters.

I bought a ping-pong ball "cage" from a bingo parlor and took out all the numbers higher than ten. I allowed those who had my loyalty card to spin the basket when they came into the store to pay for their gas, and if the last digit on their BILL was the same as the ping-pong ball . . . they "won"! They did not have to pay ANYTHING for their gas. In other words, if the ping-pong ball said "5" and their gas was $14.55, they paid NOTHING that day.

It was amazing to me how many people loved the idea they *might* win a free tank of gas. They certainly didn't want me to spin the cage—they *always* wanted to do the spinning themselves.

For the most part, this idea was profitable for me. There was one time, however, when I ended up giving away more than $80 worth of gasoline to a man driving an RV! So much for that month's fuel profits.

This idea was great for a fairly long time, but then along came the idea of pay-at-the-pump. Initially gas station and store owners liked this idea, but then they began to realize that people who used pay-at-the-pump weren't coming into the store and therefore, they weren't buying in-store items.

There have been major promotional efforts to lure the customers back into the store!

For a while, I gave away free coffee to pay-at-the-pump customers. That worked fairly well. I never did invest, however, in video equipment that might show specific items or offer specific deals to fuel customers.

By the way, Wyoming has no gambling. It does have pull tabs for charitable organizations. The ping-pong game was not perceived as a "chance" game—but rather, as a discounting technique on a retail price.

I could have set up a pull-tab opportunity, and benefited financially as the provider of that "service," but I would have had to hire people solely to run that operation, and at that point, there was little to no "benefit" financially.

People loved all these promotional gimmicks, even if they never won. They enjoyed the *chance* to win.

Billboards Were My Best Advertising! I tried the Yellow Pages and television ads, but billboards were the most effective means of advertising for me.

It makes sense, really. If you want to pull customers off a major interstate, you need to advertise where those people can see you . . . on the interstate.

It seems to me from observation through the years that most interstate highways have been built with a forecast that if the interstate is positioned just a few miles from a city, the city will grow in the direction of the interstate and eventually, perhaps decades later, the interstate will run *through* the city. And all along the way, more and more businesses will be established that give access to or lead people from the interstate.

Cheyenne is one of the few cities where that hasn't happened. Instead of growing south toward the interstate, Cheyenne has grown to the north!

If a person was driving up from Denver to Yellowstone, the "logical" place to stop would have been Cheyenne (by looking at a map). But to drive on into Cheyenne and then back out to the interstate meant an addition of at least forty minutes to the total drive time of their trip. Plus, there was only one exit to Cheyenne, and by the time a person realized they were on I-80 headed west toward Yellowstone (which is the way they wanted to go), the next exit was . . . drum roll, please . . . BUFORD! Even to turn around to go BACK to Cheyenne, they would likely be exiting at Buford.

I could never have planned such a fortuitous location!

There were only four or five highway exits between Cheyenne and Laramie. They were ranch exits—with ranchland or government land on either side of the interstate. There was no frontage area available for people to build retail outlets. The ranch owners had no desire to sell any of their land. In many cases, these ranches had been held by a family for generations.

Plus, there was no real impetus in Wyoming to develop rural areas. The entire state of Wyoming has only about 500,000 residents, and most of those people live in Cheyenne, Laramie, or Casper.

All of these factors were also to my advantage, of course.

I initially made each of my billboards different—different colors, different type styles, different messages. I figured that would be more eye-catching. If a person didn't respond to one billboard, they might respond to another one.

Then I began to notice that the Little America hotel chain, one hotel located in Cheyenne, had four billboards and ALL of them were nearly identical. A few words might be different, but the general design was the same. I learned the principle of *reinforcement.*

I redesigned my billboards and kept them pretty much the same, with a few word differences.

When the *National Enquirer* first contacted me about doing an article, I thought they perhaps were going to want to take a look inside the Old Schoolhouse or the building I was using as a garage—perhaps in search of a spacecraft or evidence of aliens from outer space. They turned out to be very nice people, who did a nice article.

When I showed that issue of the *National Enquirer* to some of my friends, they said, "That's a great photo they took. You should use it in advertising."

I had never thought about using my own photo in any form of advertising or promotion, but I mulled over their suggestion and ended up using a very similar shot on two billboards! I was afraid that people would think I was being too self-promotional, but in the end, most people seemed to enjoy meeting the guy on the billboard and they saw it as effective, fun advertising. They'd greet me as if they *knew* me—even though they only saw my image for a few seconds as they traveled by the sign at seventy-five miles per hour.

A customer told me, "Hey, you're in the *National Enquirer*! I saw you in a copy I picked up at the grocery store." That evening after I closed the store, I went to the grocery store to check out that rumor, and immediately bought up all the copies!

It was after the *National Enquirer* brought attention to Buford that I got a call from the *Today* show. The crew spent all day with me in Buford and they did an excellent little feature clip about the Buford Trading Post and my being "population 1" of my own town.

My "celebrity status" picked up after that. Not that my

celebrity had to do very much with who I am as a person or businessman. I was a celebrity because I was "the guy on the billboard."

Trust me on this; by the time I sold the store, I was very tired of looking at myself on a billboard.

Challenge #2: Get Them Into the Store

The second and greater challenge was to get a potential gas-buying customer into the store.

A truckload of fuel cost me about $25,000. My markup allowed me to make about $3,000 on a load.

On the other hand, a customer who came into the store tended to buy something, and on what they purchased, I'd usually make that same amount since most items were keystoned (a hundred percent markup; in other words, a fifty-cent wholesale price converted to a one-dollar retail price). On some items, I could mark up items to make a HIGH percentage of profit—not necessarily high dollars, but a high percentage.

I sometimes was slow in replacing the receipt paper in the automatic pumps to get the people to come into the store for a receipt.

Years later when I rebuilt the store, I purposefully put the restroom entrances OUTSIDE the store.

At times, my clerks registered frustration that people didn't read the sign with the arrow that directed them to the outside restroom entrance. I said to these clerks, "Don't get upset! The person who comes inside and asks, 'Where's the restroom?' just may see in the few seconds he or she is in the store, that there are items in the store that warrant a second look *after* they've used the restroom facilities. Let's not negate the possibility of a SALE. Tell the bewildered person where the restrooms are located and then suggest, 'We hope you'll come in and look around after you've been to the restroom.'"

Challenge #3: Get Them to BUY

I discovered that getting a customer to make a purchase is an "art" that has a lot of facets to it.

First, I had to learn what the customers wanted—*and at times, demanded*.

Ice was a very big seller in Buford. There were times in the summer months when I could hardly keep enough ice in the store. I routinely tried to keep between 100 and 150 bags of ice stocked in the freezer, but that meant a continual replenishment of the stock during any given hot summer day, and at times, my icemaker just couldn't keep up. Some people purchased ten to fifteen bags at a time—that person was usually on a "group" run of some kind, purchasing ice for multiple families back at the campground.

By all means, I had to keep the icemaker running!

Milk was another important item to have for customers. I routinely drove to Cheyenne on "milk runs." Literally. I'd go directly to the dairy and buy fifty gallons at a time and haul those gallons back to Buford in my own vehicle.

Now, there was a delivery truck that drove right past the Buford Trading Post down I-80 on a daily basis. But that truck *would not* stop at Buford to deliver a mere fifty gallons—and certainly not at the same price as a large store in Cheyenne (where the driver had to unload the milk and stock the coolers). It simply "wasn't worth their time" for the milk delivery truckers to make that stop for such a limited quantity.

But . . . I had customers who wanted milk.

What's a store owner to do? Well, a store owner does what I did—become his or her own delivery service for products that are in *demand*. And attempt to find a way to do it fast, and as inexpensively as possible.

As an aside, the entire time I operated the store I felt as if I faced an ongoing challenge of juggling between the value of my time and energy, and the amount I could make. I continually was looking for ways to earn more in any given day. That meant recognizing customer needs and doing whatever I could to meet them.

Time was a huge commodity to a store owner like me.

It is also an expensive commodity for people making deliveries to stores like mine.

I had a number of vendors in Laramie tell me that it wasn't worth their time to drive over the mountain to deliver to me, even at an elevated price. The result? There were products I would have *liked* to have carried in the store that I could *not* carry because I couldn't find a supplier to deliver the product and I didn't have the time to go fetch it.

For years, I carried Hostess pies and cupcakes. When Hostess had a sale—three items for ninety-nine cents—I made good money. Travelers loved those products. But then Hostess decided to drop its delivery service from Cheyenne to Laramie, which had taken their trucks right by my store. For me to have continued to carry Hostess items, I would have needed to go pick them up, basically paying outright for them with no return possible should the expiration date on the product occur before a sale did. It wasn't worth MY time and hassle to absorb the cost of items that were behind the expiration date, and to make the trip over the mountain into Laramie for cupcakes. Too bad.

I learned early on that if a person had to drive into Cheyenne to get one thing they needed or wanted, they'd buy everything else they needed or wanted in Cheyenne. I needed to carry enough of the basics to warrant a stop at Buford, *and* the willingness to pay a few pennies more per item.

There was no way of getting around the "few pennies" extra. Buford items always cost more at the wholesale end for one reason: freight. A store owner simply can't get an in-town price if the vendor has to travel an extra twenty-five miles to make a delivery!

I simply didn't have the volume of customers to ever entertain the concept of a "loss leader"—an item a retailer is willing to sell at a slight loss in order to attract customers. The point of getting customers into my store, in my opinion, was to make money—not to lose money!

I also had to learn that just because customers want something in one decade, they might not want it in the next!

Needs and styles and tastes changed over the years, and I needed to go with that flow.

At one time, I sold a lot of movies and DVDs, especially children's programming—and especially after vehicles began to have DVD players, or portable DVD players could be charged up and used at campsites. I sold a lot of books on tape. These items tended to be supplied by vendors who stocked them periodically with up-to-date titles.

I sold a lot of cell-phone chargers. It was interesting to me that so many people seemed to leave home with their cell phones but without their chargers.

People tended to want to buy *USA Today*, but other newspapers and magazines generally had very low sales.

Second, I had to learn where to position items. My father repeatedly gave me this piece of good advice: "If you don't know what's going on or how to do something, you'd better HIRE somebody who does know!"

That advice was put to good use at Buford.

I hired a man who was an expert in how to design a store for maximum sales. I learned from him that a "convenience" store should NOT necessarily be CONVENIENT.

He taught me basic but vital lessons about where to place certain items. The most popular and sought-after items shouldn't be close to the front door or cash register, but rather, at the back of the store. This meant that a customer had to walk past lots of other enticing items to get to the item he or she had come into the store to get. Which increased the probability of the sale of those enticing items. It had never dawned on me before that there was a financial reason why milk and eggs were at the BACK of a grocery store, but suddenly, lots of things made good sense from a marketing and sales perspective!

The previous owners had put their soda and grocery

items, mostly snack items, at the front of the store. The eating area was to one side.

Initially, I arranged my store the same way they had. Food and beverage items up front, café to one side.

After this expert's visit, we moved all of the food and beverage items to a back-room area, and relocated all of the gift and souvenir items up front—mugs, T-shirts, key chains, and a wide variety of other trinket items.

It was an expert who also kept me from making a mistake when it came to franchising the store's food services. I hired a man who was an expert in franchises to tell me if there were any of the national food chains—such as a major burger or sandwich chain—that might be interested in setting up an operation at Buford. His research took a look at a number of factors, including volume of customers, suitability of location, cost of getting basic supplies to the outlet, and so forth. There was only one major chain that even remotely qualified for location at Buford, but even that chain declined to pursue Buford as an outlet. Initially, this chain would have allowed me to purchase and run a franchise, but even that fell through. The reason? I wasn't prepared to ensure that I would have two employees on site just to run the franchise food service! I often didn't have two employees in the entire store, including myself as one of those employees.

I never had three people working at the same time in the store, except during the summer months when I ran the café and hired a cook.

Plus, there was the seasonality to the operation, which meant that a food-service franchise might be a good idea in July, but a terrible idea in January.

Challenge #4: Set Limits on "Pleasing"

In the beginning years, I had hopes that I could please every customer. As time went by, I realized—as I feel sure is true for most retail store owners—that was *not* possible.

I always found the spectrum of customers to be very interesting.

There are many customers who are happy with just about any degree of food, beverage, souvenirs, or the availability of something they need or want, and they never even question the price of an item. There are other customers who were not going to be satisfied by *anything*, at any time, and very often in a place like the Buford Trading Post they were going to be dissatisfied with the price.

I was accused a few times of price gouging, especially in times when fuel prices everywhere were fluctuating widely and moving ever higher. I discovered that the vast majority of the people who came into the store had no clue about what it takes to run a small business, and certainly not a small business smack dab in the middle of nowhere. They had no idea *why* a price for gasoline in a city might be a few pennies a gallon lower than the price I had to charge to make even a bare minimum profit—the concept that the distributors of gasoline and diesel had extra expenses for hauling fuel thirty miles from Cheyenne or twenty-plus miles over the summit from Laramie, and that those expenses were going to be passed along to me.

At other times, gasoline had been one price in their home

state when they left on a trip, and by the time they got to Buford, those prices "back at home" may have risen twenty or thirty cents a gallon—but they had no firsthand knowledge of that and saw the prices on my pumps as being outrageously high.

Still other people seemed to think that a "little store" should have "little prices." These customers didn't ever seem to think that the great outdoors, without civilization as far as their eyes could see, WAS my overhead! And a very expensive overhead at that.

Everything that came to the Buford Trading Post came by truck, and trucks take fuel, and fuel prices are automatically tacked on to wholesale prices, and therefore, most of that cost is passed on to retail pricing.

What was true for gasoline and diesel that showed up by truck was also going to be true for soft drinks and all other goods.

Is the customer always right? I have come to think *no.*

It seems to me that during the last few decades, we have become immersed in the idea that the customer is always right. Pleasing the customer is the essence of "customer service."

In truth, there are times when the customer is *not* right. But what's a store owner to do?

At the time I purchased the store, I was shocked at people who questioned my prices. I had never thought of complaining to a clerk or store owner about the price being charged for an item. I simply thought, *Do I want to pay this for the item, or not?* If I didn't, I put the item back on the shelf and chose another item or no item at all.

Sometimes I'd explain why my prices might be higher, but

most of the time, I let the matter slide by, agreeing with them, "Yes, prices are a bit higher when a truck has to haul this item all the way to Buford." Most of the time, I shrugged off the criticism I saw on some faces and let the matter slide away. The longer I was in the business, the easier it was to deal with grumpy customers and not beat myself up for not turning all customers into happy purchasers. I let criticisms roll off my back. I refused to let the complainers give me a bad day and I determined not to confront them in a way that gave *them* a bad day.

And while we're talking profit . . . As you might imagine, I did NOT like pre-priced products. They were limiting my profit margin without any input from me. A sample of this? Chewing gum. One company charged ME fifteen cents for a pack of gum and the preprinted sale price was twenty cents. Well, I needed to make more than five cents on the sale of a pack of gum.

Taking a Risk as to What MIGHT Sell

Many of the people who came into the store seemed to be surprised about two things:

- The number of items I had in the store
- The variety of gift items offered at the Buford Trading Post

My shelves were crammed with product!

Several people told me, "We don't have any 'gift shops' in our area anymore." This was stated by both Americans and people from abroad.

Through the years, I sold a number of items that I *never* in my wildest imagination thought I would sell at the Buford

Trading Post, or *could* sell. They were "not on my radar," as the saying goes.

Initially, I made most of my purchasing decisions of goods based upon whether I liked something or didn't like it. I asked myself, "Would I buy this?" If I said "no," I didn't carry the item. I discovered that what I thought was "cute" or "interesting" isn't always what the average traveler along I-80 thought was cute or interesting!

The personal preferences of Don Sammons were not a reliable gauge when it came to wise decisions in the retail world of Buford!

I had to learn to take a risk when it came to what MIGHT sell.

One man convinced me to try selling art. Let me quickly assure you that these paintings were not Elvis on velvet. They were good quality oil paintings of landscapes, birds, and animals that were native to the Wyoming region. They were framed in quality oak frames and at one time, I had about twenty such paintings hanging on the back wall.

The distributor who suggested I carry these paintings was a man who had placed quite a bit of artwork at the Sam's Club in Cheyenne. When he showed me a couple of paintings and said, "I think these would sell here, Don," I said, "Well . . . I don't know."

Oil paintings at the Buford Trading Post? I couldn't imagine tourists buying fairly large and fairly expensive pieces of art and hauling them back home—much less buying them and hauling them to their vacation destination and *then* back home. I said, "But . . . maybe."

He offered a deal in which he would take back whatever hadn't sold by the summer's end, so I really had nothing to lose.

Those paintings almost flew off the wall. What did I know?

Another man suggested that I sell mantle clocks—all different kinds. I carried a mantle clock that looked like a dolphin, some that looked as if they were straight out of elegant homes in Europe, and others of various designs. Again, this vendor agreed to swap out ones that didn't sell, so again, I didn't have much risk. The mantel clocks *sold well!*

I also sold high-end turquoise jewelry—items that cost several hundred dollars.

The man who convinced me to sell this jewelry came up from the Southwest once a year on a selling trip. He carried some amazing pieces, many of which he figured would sell in Jackson, Wyoming—fairly well known in the region for high-dollar items to those who came to see the Grand Tetons, ski the Jackson area, or visit Yellowstone.

I had a few pieces that I priced at $499—which, you must admit, is a hefty price for an item found at a place that sells diesel, chewing gum, postcards, chips, and soft drinks. Customers enjoyed looking at the pieces, but sometimes balked a bit at the price. I'd say, "Are you on your way to Jackson?" Many said, "Yes."

I'd reply, "Well, check out what you'd pay for this *there*, and then stop on your way back home and buy from me." They'd smile, nod, and walk out of the store. But sure enough, I'd see a number of them a few days later, eager to buy! The same, or a very similar item of jewelry, was selling in Jackson for $899! Now they were getting a huge bargain and they could hardly wait for me to ring up their purchase.

If a high-end item, such as the turquoise jewelry, didn't sell after a season or two, I'd donate it to the Cheyenne Symphony silent auction so I could get a tax-donation receipt for it.

Developing the BUFORD Brand

Very quickly, I recognized that many people who came into the store had a high level of affinity for the name Buford. It seemed to me at times that EVERYBODY knows a "Buford." People would routinely tell me:

- I had a grandfather named Buford.
- I had a dog named Buford.
- I called my first automobile Buford.
- I have an uncle's cousin twice removed named Buford.
- I always wanted a ___________ named Buford.

People seemed to have a strong emotional connection to the name Buford. It was a name easy to remember. A name that was a little homespun and comfortable. And all that seemed to translate to sales in the Buford Trading Post.

I began to put the Buford name on products, making them one-of-a-kind items, and sales soared for those items. I carried a wide variety of souvenir items in the store, but the most popular ones turned out to be those that had the name BUFORD on them, whether a hat, T-shirt, or mug.

I learned quickly that if you have a one-of-a-kind item, something a traveler can't get any other place on the planet, then you can charge a little more for that item—or in some cases, a lot more.

I had fun designing many of the items I carried. I certainly don't consider myself to be an artist. On the other hand, I seemed to have a knack for knowing what would sell Buford.

A salesman once advised me that if I lowered the price on my Buford T-shirts, from $19.95 to $9.95, I'd sell more T-shirts. Well, not necessarily. I knew that "volume" wasn't ever going to be a huge issue at Buford. There were only so many customers who were going to frequent the store, and only a limited percentage of them were going to buy a T-shirt. Cost was not going to be the deciding factor for those T-shirt customers. Rather, uniqueness of product was going to be the deciding factor. And people do pay more for UNIQUE.

Buford Postcards. I carried postcards of the Lincoln statue, the lone tree, and the pyramid monument. I also had postcards made of the Buford Trading Post.

The Buford postcards sold for ninety-four cents each! They were by far the most popular souvenir item I carried in the store. And with a great profit margin. I bought the cards for six cents each. And I sold the stamps for them at fifty cents each. People didn't seem to notice that the other postcards were only twenty-five cents each. They wanted BUFORD. They often bought ten or more at a time.

Most people never questioned the cost. The few who did got this reply from me, "It's the only place on the planet where you can buy this card." Nobody ever declined to purchase once they heard that.

I bought the cards in lots of ten thousand, and when I

sold, I still had about a thousand of them. I consider them collectibles at this point!

Up until the day I sold the store, I offered postcards of the old log store (the one before I rebuilt the store), and some people preferred that card even though it wasn't the store in which they were purchasing the card.

T-Shirts. A graphic artist lived about seven miles down the road, next to the schoolhouse. He was a motorcycle guy, and he had a silk-screen press that he used mostly for making T-shirts for the motorcycle crowd. He became *my* T-shirt manufacturer—I was glad to keep the business local. And I was glad to find an artist who would take my *ideas* for designs and make them even better.

I even designed a T-shirt that I was hoping the space shuttle crew would wear in space. It had an aerial view of Buford and the road sign and the space shuttle flying over "the nation's smallest town." In addition to the T-shirts, I also put that phrase on hoodies and sweatshirts.

Unfortunately, the space shuttle program ended before there was an opportunity for a crew to wear the shirt, but I left a batch of them at the launch site just in case.

On one of my trips to Florida, I took seven T-shirts, and some Buford mugs and caps for those working in management at Cape Canaveral. I paid a visit to NASA personnel there and was given a very nice tour. My hope was that the astronauts would find them unique enough to wear in space. With certainty, I could claim that one hundred percent of the people in Buford "backed" the space program enthusiastically, and that even though the astronauts had a view of the entire world, they also had their eyes on the smallest towns and villages—including THE smallest town in the United States.

I received a nice thank-you letter telling me the T-shirts were forwarded on to Houston.

Who knows? One of these days we might see someone at the international space station wearing a Buford shirt!

8

CUSTOMERS

COMMUTERS, TOURISTS, AND OTHERS

I love to tease people. And I especially loved to tease highly gullible tourists.

We had a series of snow fences near the store. They are used, of course, to inhibit drifting of snow out onto the highway. A tourist once asked me, "What are those?" as he pointed to the snow fences.

"Those are the bleachers," I said.

"Bleachers?" he asked. "What for?"

"For the llamas," I said. "They like to sit there and watch the cars go by."

He left the store with a puzzled look on his face—he actually seemed to be contemplating whether I had told him the truth!

Before I moved to Buford, I had no concept of snow fences. They are amazing—they really *do* work if they are installed properly.

The interstate was a little like a river—bringing me new people to tease every day!

During the late spring, summer, and early weeks of autumn, I could expect an average of a thousand people a day to stop by the Buford Trading Post. People might think that being out on I-80 in the middle of nowhere made for a lonely existence. Trust me, I was never lonely during those times of year. And frankly, I'm not of a temperament to get lonely. There was always something to do, fix, or install. There was always something I could think about, read about, or learn about. And if I had any tendency toward boredom, I could always go somewhere else for a few days or weeks!

One thing quickly became apparent: A place like the Buford Trading Post gives you a big, BIG window on the culture of our nation, and perhaps of people as a whole.

At their best, interstate travelers are interesting. At their worst, they can be annoying and even dangerous. But all things considered, that's no different than all of life or the business of "doing business" on any street in any city anywhere.

Tourists and More Tourists

There were two state parks and a national forest close to Buford. Regional tourists came for "recreation" to the general Buford area, and these were people who were likely to be camping or preparing for a picnic. They needed ice and snack items. And many of them stopped by to pick up a case of beer.

During hunting season, I saw an increase in customers who were headed for a hunting trip in the areas opened to hunting. I didn't carry ammunition, but I did carry hunter-oriented T-shirts, and also the spices that hunters seemed to want for making jerky from the venison they "harvested" on their trip.

In the summer, there were some people who were fishing.

I carried a little bit of line, some hooks, and a little bait.

There were also a number of tourists who were headed for Yellowstone National Park—miles away, but nonetheless, a destination point for quite a few people traveling I-80 in the summer months. We had many people stop by who had flown into Denver, rented a car, and Buford was often their first stop as they headed toward Jackson, about 350 miles away. I was 120 miles from Denver—a little more than two hours for most people coming out of the airport and finding their way across the city and northward. By the time they got to Buford they were looking for more coffee, a restroom, and perhaps a little food for the remaining trek.

In the winter months, if the snow was good at Steamboat Springs, I might have a good flow of tourists coming from Wisconsin and Michigan on their way to the slopes.

And of course, there were tourists who were making a cross-country trip from the East Coast to California. They found I-80 the most interesting and fastest way to get to California and back without a lot of desert heat.

I learned early on that most long-distance travelers seem to have an innate opinion that all of the territory between their home and the place to which they are traveling is just like the area in which they live. Many long-distance travelers have never been hundreds of miles on an interstate like I-80. They just haven't traveled that much, or perhaps they have done most of their traveling by air and not by car.

I had a number of customers who came to the store with Florida license plates. They had *not* counted on weather changes between Florida and Seattle! And they certainly hadn't counted on the wind at Buford. They had left home not remotely considering the fact that they were likely to encounter colder weather on their trip—certainly not factoring in the wind at Buford or the 8,000-foot elevation!

They'd run into the store shivering from the cold. I'd often ask, "Don't you have a coat?" They'd often respond, "I didn't bring anything like that. I'm going to California—it's warm *there*!"

I sometimes asked, "What would you do if your car broke down on I-80? If your heater stops working, you're going to be *cold*." These were questions they had never considered.

In most cases, most of the people who came into the store were very nice.

I also found that most of them had very strong opinions—about *something*, and often that something was Buford. Some were eager to share their opinions, others not so much.

Most of the people seemed happy with their personal lives, but especially in the later years, many of them were NOT happy about the state of America as a whole. Those who wanted to talk politics tended to have VERY strong opinions, and if the person discovered that I didn't particularly agree with their opinion, a few stormed out of the store in anger. I was always amazed that a person would feel it acceptable or normal to exhibit such extreme anger from a few lines of conversation with a total stranger!

Tourists from other nations tended to show up when the exchange rate with the U.S. dollar was in their favor. One year, I might have mostly foreign tourists from Europe. The next year, I might have mostly visitors from China or Japan. I could pretty much tell the state of the U.S. dollar and its valuation on the world market, and the prosperity of other parts of the world, by the tourists who drove I-80.

In Awe . . . It seemed to me that most of the tourists who were traveling I-80—either from highly populated areas of the United States or from other nations—were in awe, to a degree, of two things they found in Wyoming:

- Plenty of space to roam
- The freedom to roam

Those two perspectives probably extended to all of the Great Northwest, and perhaps even to what they would call "The Wild West" of movie and novel lore.

I had a number of tourists who were repeat visitors. The people from Switzerland were notorious repeat customers. They tended to fly into Denver and then travel by car north to I-80 on their way to Yellowstone, to the Northwest (Seattle, Portland, and so forth), and to northern California and southern Oregon. They enjoyed coming year after year to explore another part of the West!

Many of the international tourists voiced their amazement at the enormous landmass of the United States. The same for visitors from New England. I heard repeatedly, "I have been astonished at how BIG America is."

I felt very privileged that I had an opportunity to meet so many people from so many different nations. I doubt if that would have happened at a store in the downtown section of a major city.

Most of the tourists registered respect and admiration for

the United States—especially the opportunity in the U.S.A. to travel freely and to make personal life choices, purchase property, and establish businesses. I talked with several people through the years who seemed surprised that they didn't have to stop at major checkpoints in traveling from Colorado into Wyoming. Their experience as Europeans was very different.

Others registered surprise that there was so much land without houses or farms—the Buford part of Wyoming seemed extremely spacious and unpopulated.

May I Take Your Photo?

Most of my business came from the east, headed west—or headed north from Denver and then west. I had one billboard located on I-25 as travelers crossed the Colorado-Wyoming border, and another billboard at the exit where the Buford Trading Post was located. Both of these ended up with my photo on them. Because of that, in the latter years, a significant percentage of people who came into the store seemed to want to take a photo of me or with me.

After I appeared on the *Today* show in 2010, I heard from one family who informed me that they were making a trip to the West Coast and that they planned to stop at the store so their two sons could meet me. I sent them an e-mail reply that the dates they were planning to stop by would be dates that I was not going to be at the store. I was sorry to disappoint them. I gave them the dates when I would be at the store. Within a day I heard that they were changing the dates of their vacation because their two boys *really* wanted to meet the man who owned his own town.

We had a nice visit, took photos, and later had contact when I sold the town. Celebrity at such a personal level, and with young people, was celebrity that I enjoyed!

I was always a little surprised when people wanted a photo with me, or when they asked for an autograph. I realized that I represented "uniqueness" to them. That was the only way I could accept the celebrity status they gave me.

By the way, if you look for me on Google photos, you'll find thousands of photos of me. It's amazing to me how many people have posted my photo, usually with them in the photo *with* me.

Travelers of All "Makes and Models"

Over the years, I came to expect certain groups of people to stop by the store every summer.

The Hell's Angels came by every year, at least once. They were always very polite, and they never balked at prices or refused to pay. These bikers tended to ride in groups of fifteen to twenty people, and the groups were always in touch with one another. I knew that if any one group ever ran into any trouble, a lot of other "Angels" could be at that location very quickly. Over the years, I got on a first-name basis with some members of this group—they were interesting men with interesting life stories.

The Oscar Meyer Weiner Wagon stopped by one day, and we took photographs. That was a fun experience—for me, and I think for them as well.

The folks who went to the "Burning Man" event out in the desert also stopped by the store. Most of these people were artists, and many times their vehicles or trailers were outrageously decorated. At their annual meeting in the desert they would show off their creativity to others gathered there, and then as they left to go home, they'd torch their creations and leave behind only ashes in the wind. They didn't burn their vehicles, of course, but pretty much everything else. They always made a colorful addition to my parking lot when they stopped for gasoline or to get sundries from the store.

The "Rainbow People" were also an interesting lot. I never knew what to expect from them.

These were 1960s hippies. They drove old cars and VW buses, and often had tie-dyed clothes and long hair. From the smell on their clothes, it was obvious that many of them were also pot smokers. They were among my most "mellow" customers. They often sat down in the parking lot for little meetings. One time they got out chalk and wrote in the parking lot, "We love the man in the store."

Overall, they were the most unkempt customers who ever visited Buford. I was always a little amazed at who, or what, might come out of their vehicles. One time I watched two gals, two guys, and five dogs get out of a little van—the dogs looked better groomed than the people! I had to do a double take to make sure what was human and what was dog.

One day, a woman got out of a Rainbow-People vehicle. She was wearing a see-through slip. When I first looked out the window and saw her coming, I thought, *That gal doesn't have on any clothes!* There was nothing left to anybody's imagination. She slowly walked toward the store and I started thinking *fast*.

As she opened the door to the store, I saw my own sign posted next to the door that said "No Shirt. No Shoes. No Service." Just about every public place of business has a posting of that "basic" dress code.

She poked her head in the door and said, "You aren't going to allow me to come in here, are you?"

I shook my head "no" and said, "Miss, I'm sorry but you have to have shoes on to come in here."

She cracked a smile, returned to the vehicle, and fortunately for me, didn't come back.

Famous Names Came into the Store

I had a license plate that said "1 S." It stood for Buford "1" and "S" for Sammons.

The license plate that had only the number "1" belonged to the governor of the state of Wyoming. He actually stopped by the store one morning just after I had opened at seven o'clock, but I didn't recognize him until he pulled away and I noticed his license plate.

I later saw him in a store in Cheyenne and we had a brief conversation. I didn't want him to think I had been rude by not speaking to him at length in Buford.

On another occasion, Jesse Ventura and his wife came into the store. They were on their way to Los Angeles to visit their son, traveling in an RV.

Dean Martin's daughter lives in Park City, Utah, and she stopped by the store several times. She introduced me to her two daughters and was a good customer of some of the higher-end specialty items I carried.

Some people *wanted* to be recognized. Others didn't.

Not Many Truckers

A person once said to me, "I bet you really know a lot of truckers."

Not really. Not apart from Allied Moving truckers.

I didn't cater to professional truckers in the store. The truck stops along I-80 provided what they needed. Any truckers who stopped at Buford tended to stop because of the uniqueness of Buford and products apart from diesel.

Commuters Vs. Tourists

A very clear difference between "commuters" and "tourists" emerged after a few months of my being at the store. I wanted to do business with both groups, and I needed to learn to accommodate the differences between the two groups.

Commuter customers tended to know the prices in Cheyenne. The commuters were more concerned about their time. I really had to give commuters a good reason to buy fuel or other items from *me*. I worked hard to keep Buford accessible and as an outlet that provided a quick turnaround. Commuters often told me that it was faster for them to jump off the highway and get coffee and gasoline from me, than to wait in a line at a slightly cheaper fuel outlet, where they may or may not be able to get fresh coffee within a twenty-paces radius.

Traveler customers tended to be more relaxed, and frankly, less concerned about small differences in price between what was charged in Buford and what *might* be charged elsewhere. Travelers often told me that they had anticipated they could get fuel at every off-ramp on the highway—"like we do back where I live." That wasn't the case along I-80. There were NO fuel outlets at most of the off-ramps as they crossed Wyoming.

Overall, I found long-distance travelers on I-80 to have a high level of "unpreparedness." Some of them couldn't read a map, and therefore, had no idea how far away they were from various places named on their map. Others had no idea how much fuel their vehicle held, or needed. They were accustomed to purchasing gas only when their fuel gauge showed they were very close to empty. That doesn't work in areas where gas stations are several dozen miles apart.

The local customers, of course, didn't purchase very many souvenir or gift items. The travelers did. At times, however, a company in Cheyenne would purchase T-shirts or mugs in bulk—not for their use locally but to send as gifts to *their* vendors or outlets out of state. That was always a nice surprise.

Inside the store, commuter customers seem to be very focused and in a hurry.

Traveler customers tend to "browse" more—part of the reason for their stop was a chance to stretch their legs.

Commuters always knew exactly where they were, and why they had stopped.

Travelers were sometimes in a state of confusion. The road to Cheyenne is US 30 and I-80. The road from Denver intersects with I-80 southwest of Cheyenne. There is only one exit, immediately after a person merges onto I-80 to take as an exit into Cheyenne. If a person *thinks* they want to stop at Cheyenne and they miss that exit, they are already driving AWAY from Cheyenne. By the time they get to me, they are miles down the road and don't want to turn around! The net effect? A significant percentage of travelers weren't really sure *why* they were at Buford or what Buford had to offer them.

I didn't concern myself with their "why" confusion. I stayed focused on "buy."

9

ISOLATION

THE PROS AND CONS OF "NOWHERE"

One woman who came into the Buford Trading Post to pay for her gasoline purchase said as she entered the store, "This is such a God-forsaken place, how can anybody *live* here?"

I asked her where she was from.

She said, "New Jersey."

I thought but didn't say, *That's exactly what I thought about New Jersey the first time I traveled there!*

I said, "Ma'am, I don't think there is any way I could explain to you why I live here and enjoy living here, and I suspect that there is no way that you could convince me of the reasons why you live in New Jersey and like it!"

We agreed to disagree.

Although I have joked a great deal through the years and in this book about Buford being "in the middle of nowhere," I also must admit that I really didn't mind that. Isolation has its benefits.

Many people seemed to admire the fact that I could be "population 1" in a place like Buford. They didn't necessarily want to live the life I lived, but they seemed pleased to *meet* somebody who could and did live the Buford lifestyle.

The top two questions people asked me were:

• *"Are you the '1' on the population sign?"*

One of the women who worked for me as a manager got tired of being asked, "Are you the '1'?" She finally took a photo of me, put it in a frame, and when she was running the cash register, she put out that photo that had a little note attached to it, "He's the '1'."

Once the billboards went up with my photo on them, fewer people asked if I was the "1."

• *"Do you ever get lonely?"*

The answer to the loneliness question was "no."

Loneliness is something within a person. At least that's my perspective. Some people can be lonely in a crowd. Others can be alone for long hours and in remote places and never feel lonely.

I was never lonely in Buford, although I did spend considerable time alone.

A Thousand People a Day at the Store

During the summer months, I faced a stream of a thousand people a day coming into the store. I was glad after ten hours to go to my home nearby, chill out for a couple of hours, catch up on the news or watch a video, and be "far away" from the rest of the world in my own cave.

In the winter, there were always numerous chores to do, events to attend in Cheyenne, and people to catch up with who lived in the area. Plus, the store was open every day and even though only a hundred or so people might come in, that still was a stream of people.

Loneliness was not part of the landscape.

In some ways, life after Buford has had a slight edge of loneliness to it. There are far fewer human contacts these days. On the other hand, the contacts are often more meaningful.

People around me or not, I like *myself*. I'm not bored with *me*.

Shoplifters and Thieves

Somebody once asked me if it was dangerous for a person to be in the store alone—and especially if that one person was a woman.

I never had a feeling that I was putting someone in danger. In the first place, I had installed a security system that had an emergency button connected directly to law enforcement. In the second place, I had a back room that said "PRIVATE" on the door. The room was used for storage and the furnace, but customers didn't know that. I advised my employees that if anybody asked, "Are you alone?" or asked, "Is there anybody else here?" they were to nod toward the door and say, "Don is doing some business in his office." That seemed to give a sense of security to the employee and at the same time, gave a clear impression to a customer that someone in authority was close by.

In the early years, I had to leave the cash register and go outside to read the price on the diesel pumps. If I had customers in the store, I'd holler toward that closed office door, "Hey, Jonathan, I have to go out to read the pump. Watch the front for me!" It was a little bit of deception but it worked.

I also had a security camera that connected my personal office—out in the Old Schoolhouse—to the store. One time a railway employee shoplifted an item and I saw him do it. I called into the person at the cash register on an intercom I had set up. The shoplifter looked stunned. He knew that I likely had caught him in the act of stealing. He never came in the store again alone. I didn't press the point on that day, but I was glad I did what I did so he would never want to steal from me again.

I never confronted a person about shoplifting, by the way, unless I actually saw the person shoplift the item. If I saw the person shoplifting, I would confront them right away and ask them to put the item back and leave the store. Nobody argued.

The pumps were pretty much immune to theft. A few people did turn on a pump, pump a few gallons, and then drive away without hanging up the hose (which sent a signal to the cash register). They usually only pumped a minimal amount of gas, perhaps figuring that I wouldn't call the highway patrol for such a small amount. I usually took notice of the car and which way they turned onto the highway, and I watched for it to return or to drive back the other direction on the highway. I caught a few people by calling 911 and leaving the description of the car and the direction it was headed. The highway patrolmen seemed to enjoy waiting for the car to come by, stopping them, and returning them to the store to pay for their purchase. If they couldn't pay, they'd spend the night in jail.

I remember one guy who was caught a number of miles away, and by the time the police got him back to the Trading Post, I knew that he had used more fuel than he had stolen in traveling back to the store. I thanked him for his honesty in returning to pay for the fuel! He knew that I knew, and I knew that he knew that I knew, but at that point, there was no point in belaboring the point.

Staying Alert for Shoplifters

At times, I realized that there were customers who came in that had a "potential shoplifter" look written all over their demeanor. On one occasion, I had a couple of guys come in to the store and they made me feel a little uneasy as they began to say things such as, "You really are out here in the middle of nowhere, aren't you?" or ask, "Are you always here all alone?" I didn't feel any fear, but they did put me on alert. I chose to deflect their comments with a little humor, and tried to engage them in conversation. I said, "I did have a little trouble one time, but we just buried them out back." They were a bit startled but then they saw that I was joking. It turned out they were from Chicago and had never been in such a remote area.

I suspect that some people might have thought, *A guy who has a business this far away from "help" is likely to be capable of helping himself—he probably has a gun somewhere and isn't afraid to use it.* Others may have thought, *It may take the sheriff a little time to get here, but this guy probably has a plan for making any thief wait for the sheriff to arrive.*

In the more than two decades I operated the Buford Trading Post, I had a number of shoplifting incidents. Usually all I had to say was, "About the item you have in your pocket . . . did you pick up one of those or two?" The shoplifter usually pulled the item from his or her jacket or pocket and fumbled around with an apology for not remembering to put it on the counter for purchase.

I very quickly learned that the best way to avoid shoplifting was to put more expensive items right up front by the cash register, and the same for candy, gum, postcards, and other small items that would be easy to conceal.

As the years went by, I came to a realization that I really *couldn't* read customers any better after two decades of experience than I could read them the first month in business. Shoplifters and gas thieves come in all shapes and sizes—young and old, male and female, poor looking and wealthy looking, driving nice cars and old clunkers. There was no way of predicting what people were going to purchase when they came into the store. And in many cases, I decided that I had virtually *no* ability to predict which items were going to be big sellers or no-sellers.

I realized that some people who were extremely rude had no inclination to steal. Others who were very polite might turn out to be a shoplifter.

Only Two Burglaries

I was robbed twice. Actually, "burgled." It was after hours.

I had been in Buford several years when the first experience happened. I was asleep one night and the phone rang—I answered it to hear the security alarm people tell me that they had an indication that an entry-exit alarm had been triggered at the front door of the store. I looked out the window of my living room and couldn't see anything happening, and just as I was telling them that I thought it was a false alarm, I saw the light in my office go on!

I advised them to call the sheriff, jumped into my coveralls, grabbed my .38, and ran out and around the back of the store. As I rounded my way toward the front of the store, I saw a pickup truck parked close to the entrance so I waited until the burglar came out of the store, put some stuff into the back of his truck, and then start back into the store to get another haul. I confronted him with my gun drawn and told him to get down on the ground because the sheriff had been called.

About fifteen minutes later, the guy said, "Hey, I'm leaving. It's cold down here on the ground. I'm going." He started to get up.

I said, "I wouldn't advise you to do that. You've got me nervous and there's no telling what I might do with this gun."

He stayed on the ground a little longer and then said, "I'm getting up and leaving. You'll just have to shoot me." He went to his truck, got in and headed past the fuel pumps so he could turn around and drive back to the access road that led back out to I-80. It was after he turned around that I fired the gun—trying to hit the front tires of the truck. Just after he made it through the underpass to the I-80 entrance, the sheriff showed up. I gave him a description of the truck, told him the guy was headed west, and told him that I *may* have shot out his front tires. The sheriff radioed ahead and a few miles up the road, the guy was pulled over. Sure enough, I *had* hit his tires and the rubber was gone by the time he was stopped. He was running on the tire rims, generating all kind of sparks.

He hadn't really stolen that much from the store by the time I confronted him. He had a quart of oil, some candy bars, and a CB radio. The *truck*, however, was stolen! That was the *real* trouble he was in.

I probably *should* have anticipated a burglary—the door to the store was old and had single-pane glass that didn't offer much resistance to anybody who truly wanted inside. Being far from civilization, so to speak, also made the store a little more vulnerable.

For my part, I hadn't really thought through what I might do if I did face a burglar. I wasn't sure of my rights as a property owner. I didn't want to hurt the guy I confronted, but I also didn't want him to walk off with goods stolen from me. I wasn't really frightened—but I admit I was a little confused as to what I should or shouldn't do. In the end, shooting out the tires was the best choice I could make.

The second time I was robbed I didn't get a chance to confront the burglar or more likely, burglars. I was convinced it was an "inside job," and I was pretty sure I knew who was responsible. The person knew the layout of the store and knew that I had a safe with cash in it. At the time, I had some day laborers doing some repair work for me, and I was paying them at the end of the day with cash.

The burglar came into the store with a refrigerator dolly. The store, of course, was packed with goods on fairly closely situated shelves. Nothing, however, was knocked over. They had gone with the dolly all the way to the back of the store to my office, loaded my five-hundred-pound safe onto the dolly, and left the store with it. They loaded the safe up into their truck.

The sheriff came and took fingerprints. The prints matched the person I suspected of the robbery. The sheriff, however, told me that he couldn't arrest the man unless his fingerprints were found *inside* the safe.

The safe was found several days later, down a little ravine by a road. An ID sticker on the safe confirmed that it was my safe. The burglars had not been able to open the lock, but had drilled some holes so two portions of the safe could be knocked out. Some of the cash had been pulled out. The bulk of the cash, however, was in a portion of the safe that couldn't be accessed by the holes that had been made, so the burglar got far less than he *could* have stolen.

Guarding Against Vandalism

When I rebuilt the store, I reconfigured some things for better traffic flow and security. I especially made improvements to the restrooms to make them as vandal-proof as possible. I used special wall treatments so the walls were not likely to be kicked through. I had fluorescent lighting fixtures installed—the kind used in prisons—and these were virtually impossible to break. The doors were made to accommodate more than one person at a time. The doors could not be locked. In all, the restroom entrance was well lit, and more secure. A pay phone was available and 911 calls did not require money.

Plus, I lived on the property! A house with a light on inside was a constant reminder that help was nearby *if needed*.

In the early years, I had a number of customers who

would come to the door of my home asking me to open the gas pumps for them. They simply hadn't calculated how much fuel they had in their tanks, or they hadn't calculated the impact of fierce headwinds on their gas mileage. In later years, the pay-at-the-pump feature eliminated those calls.

During severe snowstorms, a number of people came to my door for help and some of these people ended up sleeping on my living room floor. I wasn't interested in finding someone frozen to death in their car in my parking lot! And I knew that it was *more* problematic for me to let people sleep in the store, or to let them sleep in my house while I slept in the store. Those who were desperate enough to seek shelter at my doorstep were usually extremely grateful for the help.

Over the years, the highway department began closing the highway sooner rather than later. If a severe storm seemed to be coming, or if a storm began intensifying, the gates were pulled across the highway before the last exit to Cheyenne. And as those measures went into effect, I had virtually no more stranded motorists knocking at my door.

Delayed Onset Loneliness?

In my early years of living in Buford, I was married and then Jonathan was born. After my wife left and later died, Jonathan was still there with me for a number of years.

After he moved away, I was "alone," but by then, I also had become very comfortable being by myself. The years of long-haul moving probably helped in that regard. When I was on the road for days, sometimes weeks at a time, there was no steady relationship associated with the driving and moving. Conversations tended to be sporadic—some meaningful and some related to the mundane facts of life and the surrounding circumstances or environment.

This is not to say that there weren't times when I wished for more companionship, but I also had an advantage of having a steady flow of people with whom to converse and exchange information. In Buford, the "regulars"—either customers or the nearby ranchers—were ones with whom I could share periodic social events, such as special dinners or an evening in Cheyenne. At the end of a long day at the store, there wasn't much energy left over for anything other than fixing a simple dinner, putting my feet up and relaxing a little without the constant interaction with customers, and then falling into bed so I could get up early and face another long day.

NOW . . .

Since the sale of the Trading Post . . .

Since I have moved to a real city and to a real neighborhood where I have real neighbors within shouting distance . . .

I have found myself home alone all day—and I'm starting to feel a little lonely.

I have found myself looking for projects to do. I've remodeled the bathroom of my home and am working on the kitchen cabinets. And then? I'll be looking for something to *do*.

I am in search of a social network for the first time in decades.

Here I am in the middle of a populated area with no social

life—in contrast to the Buford years in which I was in a highly unpopulated area with lots of social interaction.

There are going to be some *changes* in Don Sammons' life. And they will involve *people!* Maybe not a stream of a thousand a day, but definitely I will no longer pursue the isolation of being "population 1."

10

STREAMS OF INCOME

EVERY DOLLAR HELPS

When I first took over the business, I had lots of days in the winter where I barely topped fifty dollars in gross income.

For a number of years toward the end of my time at Buford, I grossed more than a million dollars a year, all aspects of the property included. Obviously, that wasn't NET. But it certainly was a far cry from the beginning days.

The key to that kind of increase in income can be boiled down to one word: Diversification.

I mentioned in an earlier chapter that I saw business as a challenge, not unlike a sport. I always felt a certain adrenalin rush when I tried something new and had it turn out to be highly profitable. Over time, I saw that diversification was the name of the "game." It was much easier to make a living that way than to devote massive amounts of energy trying to get a few more customers into the store.

Several wonderful opportunities came my way. And I seized them.

A Cell Tower, P.O. Boxes, and a FedEx Turnaround

A cell tower company approached me about putting a cell tower on the property. It was a new company in Wyoming and its signals were transmitted on a "line of sight" basis. On I-80 they had a tower on the summit that could receive signals from Laramie, but that tower could not transmit all the way to Cheyenne. They needed a property that could provide intermediate transmission.

We negotiated over several months and in the end, I agreed to a tower that was a freestanding tower and we signed a twenty-year agreement.

I made a similar deal with the U.S. Postal Service.

After the post office was closed *inside* the store, the United States Postal Service put a bank of boxes outside the store and down a small road. The boxes were on state property. The U.S.P.S. didn't realize, however, that they had miscalculated and put one part of the little structure on a rancher's private property. They also placed them in a way that a customer had to stand *in the road* to get his mail.

Over time, some truckers who pulled off the road to sleep knocked out some of the boxes because they couldn't see them in the dark. The postal service put up a guard rail, and at that point, the rancher balked and said, "I want all of this off my property and out of here."

The postal service contacted me and asked if I had any place outside the store where they could put their boxes. I showed them an area of the parking lot that was very seldom used and they leased a little fifty-foot by four-foot strip for the mailboxes. I insisted they build a little roof over the boxes in a style that matched the store. I told them I would keep the lot plowed, but I didn't want that in our agreement. They offered me

$500 a month for the lease.

It didn't make any sense, of course. They could have put them on the other side of the fence on state property and the postal service wouldn't have had to pay anything.

I happily accepted the U.S.P.S. money. The boxes served one hundred and twenty-five people in the greater Buford area and I was also happy to see these people come into the store after they picked up their mail.

Later, Federal Express wanted to lease a "turnaround area" on my parking lot. I was reluctant but the more times they called, each time upping the price offered to me, I eventually agreed to accept their deal. It seems that Federal Express has a limit on the number of hours their drivers can work in any given twenty-four-hour period. Given the way they had set their routes in Wyoming, the trucks coming from Laramie in the west could not make it all the way into the yard in Cheyenne in the allotted time. And the trucks coming from the east couldn't make it to the yard in Laramie. But the trucks *could* make it to Buford.

Federal Express wanted to "lease" turnaround space on my parking lot on the basis of their buying fuel from me for these trucks. I didn't want that deal—it would have meant being open for the sale of fuel between one o'clock and three o'clock AM, otherwise known as the middle of the night! They finally agreed to a monthly cash amount and I accepted. I never saw those Federal Express trucks. I did see, and cash, the Federal Express checks.

These three streams of steady monthly income were not in my wildest dreams when I first took over the Trading Post.

They provided enough in later years to pay my personal bills.

A Salvage Yard Just Beyond Sight

Behind the store, down an incline, I ran a licensed salvage yard. The advantage of having a salvage yard *licensed* by the state was this: the state highway patrol could call the owner to tow vehicles that had been abandoned on the interstate or state highways. Unlicensed salvage yards with tow trucks were not called.

There's really no telling why some people abandon their vehicles. I sometimes towed fairly nice vehicles to the salvage area—nice in appearance, that is. I knew enough about auto mechanics that I could tell if a vehicle could be fixed, and the degree to which a "fix" might be a good business deal.

Let me add quickly that the vehicles that were stored for salvage or sale were not visible from I-80 so there was no problem with them being an "eyesore" or anything that might detract from the business of the store.

The abandoned vehicles that were towed to my salvage yard were usually sold at auction by the county sheriff. A salvager had charges against the sale (for towing, storage) and those were recovered at auction. If nobody bid, the tower was given a title for the vehicle and could dispose of it as he saw fit. Usually that meant selling the vehicle to a crusher for potential recycling of the metal.

Through the years, there were several vehicles that I ended up selling for a good amount of money.

One woman was traveling through the state in a vehicle

that was almost new. She had liability insurance, but that was all. When she slid off the road, there was no insurance money coming her way. She didn't want to stick around to try to dispose of the vehicle. She asked me if I would sell the vehicle for her, for a percentage, and she signed over the title to me. I made several thousand dollars on that deal.

In another case, I towed in a pickup as an abandoned vehicle. Later, I learned that the driver had left his vehicle to hitch a ride into Cheyenne to get a part needed for the repair of the truck. He had been picked up in Cheyenne by law enforcement officials who discovered that there was a warrant for his arrest from a New England state. He was put in jail and eventually sent back East, and the pickup was mine to dispose of as I saw fit, without going through the auction process. I sold it for a nice sum, grateful that it hadn't been stolen property (in which case there would have been *no* money).

In my best year, the salvage yard netted me $80,000. It was not only a good side business—it was a good business!

Sending a Telegram from Buford

I ran a towing service out of the Buford Trading Post—more on that in the next chapter.

I discovered in the towing business that many people didn't have funds on hand to pay for towing services. That meant that after towing their vehicle into Cheyenne for repair, I would need to wait for them to go to the Western Union office to wire for funds. That ate up a lot of valuable TIME. So much so that I eventually became a Western Union store. That way, people could wire for funds to be sent directly to me, and while we were both waiting for the money to transfer, I could continue to wait on other customers at the store. Once the money arrived, I would drive them into Cheyenne and drop them off.

At times, I had so much towing business that I kept three tow trucks busy. That's one driver—ME—operating three trucks on something of a relay basis. I had two wrecker-style tow trucks and one flatbed truck. On a busy breakdown day, I could drive one truck into town and jump into a truck I owned but had parked at the garage, go back out to the highway to bring in another vehicle, and by the time I returned, have the original truck unloaded and ready for the drive back to Buford. There were times when I would have two or three people waiting in the store to have their vehicles, and them personally, taken into town. I eventually purchased a "trailer" that I could pull behind a wrecker-style tow truck so I could haul TWO vehicles and their passengers into town. I saw no reason to give away my towing business to in-Cheyenne tow services simply because I didn't have enough equipment to meet the need.

And yes, there were that many breakdowns on I-80.

And yes, the business was that lucrative at that time.

Parking Spaces "By the Week"

In a grassy area adjacent to my parking lot on the *private* side of the property but not real close to my house, I put in fourteen little poles that provided electrical power for fourteen parking spaces. These spaces were not for RV users, who tended to want, and needed, more services than I was willing to offer.

These "spaces" were usually leased to workers employed by the railway, various pipeline-construction companies, a nearby quarry, or the state highway system (as pavers and repairmen). The workers had their own pull-trailer campers for overnight sleeping and storage of clothing and basic sundries. These workers weren't looking for any form of luxury or amenities—they *were* looking for cheap lodging close to where they were working.

I charged them for electricity only, $10 a night or $50 a week if paid in advance. You can do the math. Have all fourteen spaces rented out resulted in $700 a week, or more. That's not a bad stream of income. It more than paid my electric bill.

I let them use my garden hose for water, but offered no dumping or sewer services. Some of them used my restrooms at the store. Many of them used the showers at a truck stop down the road. And in the evenings before I closed, many of them came into the store to get cigarettes, beer, or to order a pizza that I had purchased in bulk and could readily pull from my freezer and heat up for them in my pizza oven.

On one occasion, the guys who were working as pavers for the highway system told me that they had ordered way too much paving tar, and I gave them the opportunity to solve their problem by paving my gravel driveway! A paved driveway for a couple cases of beer is not a bad deal.

I also benefited from excess asphalt.

Some of this was dumped in my salvage area with EPA approval and it served as excellent landfill that made access to the salvaged vehicles easier.

Every Dollar Counts . . . And Helps

I am a strong believer that no amount of money is "too little"—especially if it comes in regularly and reliably. I still stoop down to pick up a penny from a sidewalk.

These extra streams of income were very important to me, and the best news about them was that I neither needed to ardently solicit them, nor did they take a great deal of time and energy. In the case of the spaces for state workers, the income came with a certain degree of camaraderie.

In the case of the Federal Express turnaround contract, the income came without any human contact!

Both methods worked for me.

11

THE SHOOTOUT

THE TRADING POST PARKING LOT BECOMES THE OK CORRAL

Shortly after I moved to the Buford area I recognized that a fairly large number of vehicles became stranded for one reason or another along I-80, especially west of Buford on the way to Laramie. Some of these vehicles needed towing to a place where they could be repaired. Other vehicles were simply abandoned.

Once I started the Buford Trading Post, these two "needs" for towing and disposal became an interrelated business opportunity!

For the salvage yard, of course, I needed a vehicle to get abandoned vehicles to the salvage area. For those who were "stranded," I needed a tow truck to get them to Cheyenne for repair.

I invested in a tow truck, and a snowplow that could be added to the tow truck if need be in winter months. I was approached by AAA about the possibility that I might work with them as a AAA service provider. The previous Trading Post owner had been a AAA provider so it seemed like a potentially good deal for me to drive on their behalf as I had once driven a rig for Allied. I explored the deal with AAA and learned that they set the rates for towing services, regardless of whether I could provide those services for that amount. I decided at that point that I did *not* want to be a AAA tower. I opted to have an independent towing service.

The state highway patrol knew Buford, of course, and once I had the towing service, they knew that I was the closest service provider for the majority of those who became stranded between Buford and the summit (about ten miles west) and even over the summit toward Laramie. This put me, of course, in direct *competition* with AAA.

A problematic situation arose when I had a AAA member come into my store needing a tow. They'd call AAA in Cheyenne and then have to wait upwards to an hour for a tow truck to arrive.

Meanwhile, they wanted to wait inside my store for that tower to come to their aid, and invariably, this seemed to happen repeatedly at *closing time*. The AAA advised me to put them out in the cold and close up!

I told them I'd provide warmth and shelter for their customers, but that I needed to be reimbursed for overtime pay to my employees who kept the store open for them. I sent them an invoice to that effect . . . which they refused to pay. And, they sued me to keep from paying the invoice. I won about ninety percent of that suit, but in the end, the problem wasn't solved. The case went all the way to the Wyoming Supreme Court.

What I learned in that process was that AAA *did* have the right to come pick up their customers who had a breakdown, but I also learned that the AAA tower only had the right to come onto my property—in other words, the parking lot—to pick up a AAA member. The AAA tower did not have the right to solicit business on my property, or to fix automobiles on my property.

I ended up in a longstanding feud with AAA. At the core of the feud was basic business competition . . . that went amuck.

At the heart of the matter was that there was no place designated by AAA for their tow-truck drivers to do simple repairs, such as changing a tire, replacing a battery, and so forth. The highway was not constructed in a way that had turn-outs or a wide enough shoulder for these repairs to be made where a vehicle had become inoperable. The vehicles needed to be towed to a safe flat area. The parking lot of the Buford Trading Post *appeared* to be the perfect spot.

Never mind that this was my private property.

Never mind that the gravel parking lot was *not* an ecologically ideal place to fix any kind of problem related to oil, gas, or any kind of problem related to fuel hoses.

Never mind that the tow trucks tended to be a nuisance to regular customers. They frequently blocked access to the fuel pumps, post office boxes, or the front door of the store.

Never mind that the appearance of ongoing repairs made the Buford Trading Post look like a mechanics shop instead of a place to buy ice, snacks, and souvenirs.

The truth is, I didn't use my own parking lot to *make* repairs. I towed people into Cheyenne. If I didn't see good marketing or business advantage in making repairs on the parking lot outside the store's front door, why would AAA think I would approve of *their* making repairs on my lot?

The AAA guy who seemed to have been assigned my

stretch of I-80 broke that general rule about repair work repeatedly, even when he was advised by both me and by his AAA supervisors that he could not and therefore should not and must not service vehicles at the Buford Trading Post.

Keep in mind that this man was a direct competitor to the towing business that I established after I decided *not* to sign a contract with AAA. The local sheriff would come out and give this man citations for trespassing, but never actually evict him from the property. This went on for eight very long contentious years.

One day I reached my limit. That morning I had a call from the state unemployment office about a former employee who was filing for unemployment even though he had *not* been fired but had resigned for his own reasons. I had also received a call from my former wife Debbie who was presenting me an opportunity to pay for something she believed our son needed and deserved. And THEN the AAA competitor walked into the store. I said, "What are you doing here? You aren't supposed to be on the property." He started claiming that he had rights to be in the store and on the parking lot, complete with lots of expletives.

He went back out to his truck on the parking lot, but didn't move.

I went out and told him that he needed to drive on down the road. And I added, "If you are here when I come out of the store in a few minutes, I'm going to shoot out the tires of your truck. And furthermore, every time you come on this property, I'll do my best to shoot out the tires of your truck. Perhaps eventually you will understand that you are trespassing and that you have NO rights or privileges to do AAA towing business on this parking lot."

I walked back into my office, which was on the opposite side of the store, and got a .38 special out of my safe, came back out to the lot and since he was still there, I did what I had said I would do. I shot out the tires of his truck. He managed to drive out of the parking lot, but just a few miles down the road, he was driving on rims. He called the sheriff.

And within a few minutes, my parking lot was stormed by law enforcement officials. I was ordered to come out of the store with my hands up, and after they threw me to the ground and handcuffed me, they told me to open the safe and turn over the gun. I argued that they needed a warrant for that, but when they threatened to close down the store and keep my employees from coming to work, I opened the safe. They took the gun, and they took me to the county lock-up facility. They refused to allow me to post bond, stating that I was a threat to the community! They convinced the judge that I was likely to return to Buford and shoot up the entire store.

Now, these law enforcement officials had known me for years. The ones related to I-80 patrol knew about my towing business. The sheriff knew that he had issued numerous trespassing citations to the tow driver who now needed new tires. Nevertheless, they filed serious felony charges against me and kept me locked up thirty days without any recourse for bond or an early hearing of my case.

I was in a cell with a young man who was incarcerated for tying up a gay guy to a fence. Frankly, there were moments when I thought it was all a very bad dream from which I couldn't seem to awaken.

Let me also add that the days I spent in jail were days that I consider to be the most useless days of my entire life. The only thing the prisoners did there was watch junk TV and play cards. I don't enjoy either pastime.

The woman who was helping me manage the store at that time kept the place running. She'd bring checks to the jail every week so I could sign them and keep the Buford Trading Post open. This woman, as well as some of my friends, wrote letters on my behalf—stating that I was *not* a threat to Buford and when the judge finally heard my bond hearing again, I was let out.

When the case went to trial, the judge did not allow ANY of the previous trespassing charges against the AAA to be submitted as evidence. He did throw out the "possession of a fire arm" charge because the gun was taken from the safe without due process. But, the judge *did* let the charge of "aggravated assault" stand. That's a felony charge that can result in a maximum of ten years in prison!

The jury found me guilty.

At that point, all ideas that this was a little funny flew out of my mind. I got very serious about the potential that I could spend a decade behind bars.

My friends rallied around me, and on the day of the sentencing, the entire courtroom was filled with people who were standing with me. Many of them came forward to speak positively on my behalf.

In cases of first offense—which is what this was for me—the recommendation is nearly always "probation."

The tower, his attorney, and the probation officer

assigned to me—who was a friend of the AAA tower—recommended the full ten years of imprisonment.

A jury had said I was "guilty" of the charges brought against me, but the judge alone gave the sentence. He ruled in favor of five to seven years in the state prison, but then stated that my sentence was being suspended in favor of five to seven years of supervised probation.

I requested a probation officer from Cheyenne, even though it was in another county, and that was granted. After two years, I was told that I no longer needed to report for probation. I had paid my debt to society in full.

Since then, I have been filing repeated letters to have my felony charge overturned.

I've had several legal experts write statements on my behalf stating, in part, that I was way "overcharged" for the severity of the crime I had committed in shooting out the tires of a trespassing competitor. The governor's office, however, has been highly reluctant to consider pardons for incidents that have not passed a ten-years-ago mark. That mark has now been crossed and then some, so I'm hopeful that the elections in 2014 might be my opportunity to receive a pardon and have the felony expunged from my citizenship.

It would make crossing over into Canada on a motorcycle much easier.

And, I'd really like to have my voting privileges restored.

I do know that there are still people in Laramie who talk about the "shootout at the OK Corral in Buford." Very few of them have the story straight, or complete. But it makes for a little folklore in that part of Wyoming.

12

BUDDHISM

A PHILOSOPHY THAT RODE IN ON A HARLEY

The shootout at the Buford OK Corral might not have occurred if I had met Fran before that time.

Fran Robinson rode into the Buford Trading Post one day on her Harley. It was unusual to see a Harley pull into the store parking lot ridden by anything other than a motorcycle club or the Hell's Angels. It is especially unusual to see a Harley in Buford ridden by a woman, and a pretty woman at that.

I have owned and ridden a Harley for many years, so we had an immediate connection based on our respective "hogs."

Fran told me that she was going on a spiritual retreat at a place down the road several miles. I knew about the place. A painter and his wife owned a cabin they used for periodic retreat weekends, and I had always assumed that it was a church-based center. Which is what it *mostly* was. But, Fran was going with her own spiritual agenda and she told me in plain and simple language, "I'm a Buddhist."

A pretty woman riding a Harley who is a Buddhist?

This was not your average traveler through Buford!

I began to talk with her about that—and since the store was mostly empty—we had a good opportunity to talk about motorcycles and her form of Buddhism for more than an hour. She was part of a Buddhist organization that had a center in Fort Collins, and she gave me a card with more information about the group. She invited me to consider attending one of their meetings.

A couple of weeks later I went to a meeting and found it very interesting. I wanted to know more. The more I learned, the more I felt this was a good fit for me. I made a decision that was as simple as the one Fran proclaimed, "I'm going to be a Buddhist."

I am part of a group that is located in Greeley, and she is now part of a group that meets in Windsor.

Up until the day Fran walked in the door of the store, I don't recall ever hearing anything about Buddhism—in my entire life. Granted, I served in the U.S. Army in Vietnam, a mostly Buddhist nation, but I certainly never learned anything about Buddhism when I was in the military, and I had never had any experiences with Buddhists in America.

Many books classify Buddhism as a religion, but from everything I have learned, I classify it as a philosophy of life.

The sect of Buddhism to which I belong doesn't "meditate" or do any form of yoga. There are no special garments or religious leaders. We chant—individually and collectively.

I initially found this form of Buddhism to which I now belong to be very attractive because it called me to serious personal responsibility for my own emotional responses to life. I came to recognize that no person is ever going to be "perfect" in any type of relationship. I will always be disappointed to some degree because no other person is ever going to be identical to me. I also came to recognize that no person can *force* me to have an emotional response, or to exhibit vengeful or angry behaviors. The display of behavior is always my responsibility.

I have come to believe that I have the power in my own mind to create my own environment in which I can make decisions and experience emotions that are for my personal benefit and growth as a human being. I can choose to act or react, and as I do, there is a consequence that is created immediately, even if it does not appear materially for some time. It's a little like a person smoking cigarettes for thirty years, and then stopping, and then being diagnosed with lung cancer five years later. The decisions I make today are going to "appear" in the form of some type of consequence—good or bad—at some point, usually unpredictably but nonetheless real.

No person can *make* me happy. I must choose to make myself happy!

I became aware that I do not need to respond to circumstances at the level of *instinct.* I can respond with reason and at a level of greater understanding. The Buddhist term for all this is "enlightenment." Every person can put himself or herself into a situation to gain greater "light" on a circumstance. I can view life much more objectively, and respond to it with a positive frame of mind.

Instincts are not necessarily bad—but at the same time, they are not necessarily good. Fear is an instinct that can keep a person from accomplishing great things. Or, it can protect a person from physical harm. Operating at a level above instinct gives a sense of responsibility *over* a person's emotions. That has been very beneficial to me.

Nam Myoho Renge Kyo

I started practicing Buddhism in February 2003. That was four years after the "shootout" incident in the Buford Trading Post parking lot. I feel confident that I would not have shot out my competitor's tires if I had been a Buddhist at that point.

I can go back and visually recall virtually every minute of the tire-shooting incident, including the emotions that I felt each minute. I can feel how upset I got, and every word I spoke. Those simply are not emotions that I now feel, and the words I spoke are not words I now speak.

On that day, I simply lost control of my emotions. In fact, I believe I was running on emotions alone. There was no logic, no regulator. Since that time, I have had equally disturbing circumstances occur, but none of them have ever robbed me of my reasoning and put me back into a position of acting on negative emotions alone.

Overall, I am much more peaceful with myself.

I am able to stay in a frame of mind that is much more "reasonable."

I carry a card in my wallet that has these four words on it: Nam Myoho Renge Kyo.

The words mean:

- Nam – devotion to Buddhism
- Myoho – mystic law, an essential part of life
- Renge – cause and effect
- Kyo – truth expressed through the sound of one's voice

It is a card that reminds me of my commitment to chant regularly.

The sect of Buddhism to which I belong is called Soka Gakkai International (SGI), which means a "value-creating society." There are twelve million adherents worldwide. SGI advocates a form of Japanese Buddhism based upon the teachings of Nichiren Daishonin, who lived in the twelfth century and espoused the philosophy that became known as the Lotus Sutra. One of the most famous principles is, "Your life through cause and effect." In other words, a person must become responsible for his own actions and reactions.

The current president of SGI is Daisaku Ikeda, who lives in Tokyo and is in his eighties. He has written numerous volumes of poetry and literary essays and has received honorary degrees from more than three hundred and fifty institutions of higher learning. I consider him my mentor—in other words, I am a disciple of his teachings.

The SGI form of Buddhism does not have priests, a religious hierarchy, sacrificial offerings, a devotion to yoga, or many other things that people often associate with Buddhism. The focus is on personal and corporate chanting to generate peace, both within and among people.

For me, it has been a key to anger management and to feelings of inner well-being.

For years, I traveled by motorcycle every Sunday from Buford down to Denver to spend time chanting with other SGI members—our focus was chanting for world peace.

The Only Honorary Citizens of Buford

The only time I ever gave anyone the "key" to the town of

Buford—a symbolic gesture, not a set of keys to the front door of the store or keys to the gas pumps—was to Daisaku Ikeda.

I framed a copy of the proclamation that I had delivered to Daisaku Ikeda, and kept it hanging in the store. I received back a very nice personal letter from him. He thanked me for the honor and noted that the entire town of Buford was Buddhist, not likely a claim that any other town in America could make.

Below is the declaration related to this:

GREETINGS in the name and by the authority of the town of Buford, Wyoming.

Honorary Citizenship Document

Comes now before all who are present and all who will hear of these proceedings in the future.

Let it be known to all that the owner of the Smallest Town in the Nation of the United States of America, Buford Wyoming [Population (1) one] and its only resident, Don L. Sammons who is empowered through his position as owner of this great piece of Wyoming history which was established in this wild and uninhabited part of the west in 1866:

Whereas those early pioneers worked to tame this land and bring civilization to the Wild West so do President and Mrs. Ikeda also have that pioneer spirit to work for the advancement of the Soka Gakkai International, a socially engaged, ethnically and culturally diverse Buddhist organization, with 12 million members in 192 countries and territories.

Whereas because of the outstanding work and efforts of President and Mrs. Ikeda to spread world peace through their shining example and the SGI-USA organization, Buford Wyoming wants to honor these two dedicated and committed individuals who are truly deserving of the praise and honor bestowed upon them, now therefore be it.

In Daisaku and Kanoko Ikeda we have outstanding honorary citizens, worthy of the esteem of both the Community and the great State of Wyoming.

Whereas on this date of October 20, 2009 I, Don L. Sammons, bestow on President Daisaku Ikeda and Mrs. Ikeda Honorary Citizenship with all its privileges of the town of Buford Wyoming.

In Witness Whereof, I have hereunto set my Hand and Signature.

I signed the document as Don L. Sammons, Owner and Chief Spokesman and attached my personal seal as the "Official Brand of the Town of Buford Wyoming."

13

THE FIRE

LIGHTNING CAN CHANGE EVERYTHING

BAM. BAM. BAM.

The loud and insistent knock on the door of my home came at one thirty one morning.

It was a trucker and I don't mind telling you that I was a little irritated.

When I answered the door, he said, "Hey, man, your store is on fire." I looked out and saw flames at the back of the store.

I immediately called 911, and then jumped into my coveralls and ran to the store and tried to hook up the hoses to fight the flames myself in the interim. The water system, of course, required electricity to activate the pump, and the fire had knocked out the electricity.

There was little I could do except to wait for the fire department. It took them about forty-five minutes to arrive after I placed my 911 call.

The fire did not spread quickly because the store was closed up, and was fairly tight. There wasn't much oxygen to fuel the flames.

The store was struck by lightning in August 2003. It was not a "direct hit," but rather, a lightning strike that ran along an overhead electrical wire for more than a half mile until that bolt of energy slammed into the electrical box on the outside of the store. The result was a small electrical fire that grew in intensity and size, fairly slowly but relentlessly.

It was a dry-lightning storm—which meant there was no rain to help put out any fire that might have been started that night by lightning.

The store that burned was actually the third store on that site. The original store was built in 1866 and was destroyed in a fire. The second store was built in 1895 and at the time I purchased the property, it was still a decent building. I converted it into a garage and stored my personal cars there.

The Old Schoolhouse was used for storage and for my personal office.

As I saw the flames coming out the back of the store after the lightning strike, and then noticed that the surfaces of the propane tanks nearby were beginning to "bubble," I quickly went into gear. The propane tanks had just been filled. So I knew that they could and would cause quite an explosion.

I also knew I had to get my cars out of the garage—at that time I had a four-door Jaguar sedan and a Corvette. But there was no electricity, of course, to open the garage door. I finally resorted to pulling the rope manually and dashed back and forth as quickly as I could to get both cars out of range of a propane explosion. That effort took my mind briefly off the fact that my workplace was going up in flames.

When the fire truck arrived, they immediately focused on cooling down the propane tanks.

The next fire truck to arrive on the scene was the volunteer truck. They were determined to get inside the building.

Up to this point, the fire had been starving for air.

These guys, however, were determined to get inside. They knocked down the front door to the store, creating a rush of air into the building, and the fire went crazy. Within ten minutes the roof collapsed and all they could do was shoot water into the flames.

The fire was still smoldering at eight o'clock the next morning. The walls, which were logs, were charred or in ashes.

I knew the second the firemen knocked down the front door that the fire would win. Remember, I had BEEN a firefighter in St. Louis.

Total Devastation— Materially and Emotionally

I can hardly express in words how devastated I felt after the fire. So much of my personal memorabilia was in the store—things related to my personal history and interests as well as the history of the Buford area.

Why were these things in the store and not at home? Because I spent most of my waking hours, seven days a week *at the store.* That's where I interacted with tourists, acquaintances, and in many ways, with my friends in the area. The store was like an extension of my home.

Those who came to "investigate" the fire came to their conclusion very quickly. The evidence was obvious to them. They knew it was *probably* lightning so they consulted the forestry maps and they learned that there were five thousand lightning strikes that night in the immediate Buford area—each strike could be pinpointed. They isolated a strike that readily could have hit the electrical wire that ran to the store—and such a strike was totally consistent with the damage they found to the outside electrical box on the west side of the store.

The insurance people, of course, felt compelled to conduct their own investigation and they had a guy come and go over every inch of the inside of the store looking for any evidence of fluids used to ignite or accelerate a fire. I told them I had a couple of gas cans in the storage room because I often loaned out gas cans so people could get their cars started after running out of gas along I-80. They weren't interested in old gas cans. They were looking for trails on the floor of something that might have been used in arson.

Initially, the insurance company refused to accept the report that the fire had been caused by lightning. They had an adjustor on site for several days sweeping the inside of the charred store for any evidence that the fire had been set. They sifted through all the remains looking for any sign of an accelerant or the ignition source.

I was advised by the legal authorities not to clean up *anything* inside the store. They told me bluntly, "If you claim two refrigerators and the insurance investigators don't find two refrigerators, you won't be reimbursed for them."

I discovered two major lessons in the following months.

First, the process of insurance investigation can be incredibly long—at least from the policy-owner's standpoint. The insurance adjustors weren't finished with their work until the following March, seven months after the lightning strike.

The insurance people obviously found no evidence that the fire had been set.

The second lesson—you probably never fully understand what's in an insurance policy until you begin to file a claim. And even when you discover what's covered and what isn't, the definitions of certain terms may be interpreted differently by the insurance company and the policy holder. I had "loss of income" in my policy, but I quickly discovered that the insurance company regarded that as a "net figure," not a gross figure. This meant that they were using my net-income calculations from the previous months and years to calculate payment to me. The gross figure, of course, included income that was necessary to cover expenses such as taxes, utilities, fuel, all the supplies necessary for the running of the business, and so forth. The net was what I took "home" as a small-business owner. When the insurance company gave me "net" they gave me what amounted to an employee stipend, not the amount necessary to run the business.

Furthermore, they calculated the "net" that was related to fall and winter months. The big-income months at the store were the summer months between Memorial Day and Labor Day. "Net" for those months was a decent number. The "net" for the low months of late fall and winter was an amount that often didn't stretch to cover basic living expenses.

Was this a matter of concern? Yes!

Stress? Yes!

The Decision to Rebuild

I faced a major decision: Should I rebuild?

And if I did decide to rebuild the store, how should I rebuild it?

I studied some of the trade magazines and learned that if a "canopy" was built over the fuel pumps, sales tended to increase by twenty-five percent. The big chain distributors of fuel who had studied this had determined that a canopy made customers feel safer and also raised visibility for the business. All that made sense.

I had a gravel parking lot for the old store. I decided that it would make much more sense to have a paved parking lot, which also provided a better foundation for a canopy over the fuel pumps. The previous building had about 2300 square feet. It had a kitchen, and a residential area—a bedroom and bath. When the store was originally built, the people operating it had also made it their home. I had used the bedroom for an office and storage.

There were two other changes I wanted. The old fuel pumps had not provided for credit-card purchases. All of my competitors within a fifty-mile radius had credit-card readers on their pumps so I made the decision to buy the new pumps.

The old roof had been a "log" roof. Keeping in mind that the store had burned indirectly as the result of a lightning strike, and knowing that lightning strikes were frequent in the Buford area, I also determined that it would be safer and wiser to put on a different kind of roof.

Many of the things I lost in the store were things that were no longer manufactured. Remember, the store had been built in 1939!

I did some sketches and measurements and got some estimates from a contractor. I decided that I could make the store 1550 square feet, eliminating the living area, and the difference in cost might allow me to have a canopy over the fuel pumps, buy the pumps I wanted, pave the parking lot, and put on a fire-resistant roof.

I told the insurance people that I had no desire to replace the living area that had been at the back of the store. I told them I wanted to rebuild the store square footage but NOT the living area. I asked specifically how much the living area was worth. The figure I was given was enough to put a canopy over the pumps and pave the parking lot. Both of these things were a matter of modernization and directly related to industry literature that indicated these features could boost sales. I was also advised to consolidate the fuel tanks, also as a means of modernization to allow for a third grade of gasoline. I duly copied and faxed or sent photos and invoices and all the details of what I was doing as I rebuilt the store.

As I faced the final payments owed to the contractors and was ready to receive the final check from the insurance company, THEN the insurance company said that they were deducting $150,000 from my claim because I had not *replaced* what I once had—instead, I had "upgraded."

My adjustor had been telling me one thing. The insurance company was now telling me something else. The bottom line was that I would be receiving a set amount of money, all related to "like for like." When I asked about this policy of "like for like," I was told, "Well, you had fuel pumps for gasoline and diesel—your new operation should have fuel pumps. You had a store that sold gift items and various sundries—your new operation should do that." The beef now was that I had not rebuilt the store using the exact *same* pumps, roof, and so forth.

There was yet another complicating factor.

At the time of the fire, I had a number of checks written and sitting in my office drawer, ready to be mailed at periodic dates during the coming month. Those checks went up in smoke. One of the checks was the payment for my health insurance. With so much going on, I didn't think about my health insurance not being paid.

The following month, I realized that I hadn't received an invoice for health insurance and I called the company, only to learn that my policy had been canceled! The health insurance was willing to reinstate my policy if I brought the amount owed current, and also agreed to automatic payments. A couple of months later they took out the first of the automatic payments. And I was certainly glad they did!

Two weeks later I had a major heart attack.

It happened while I was in Cheyenne one day. Friends there asked me if I wanted to go with them to the gym where they worked out. They said they could make arrangements for me to have a guest pass to see what the facilities were like, with the possibility that I might eventually join the gym. We were going to dinner after our workout time, so it was a good opportunity for exercise *and* friendship.

I cycled my way through thirty minutes on a stationary bicycle and started in on thirty minutes of treadmill walking.

About fifteen minutes of the treadmill, I felt a wave of extreme exhaustion. No pain. But incredible "tiredness." I made an assumption, *Wow, Don, you are far more out of shape than you thought you were.* I managed to finish the thirty minutes I had planned for the treadmill, and went to take a shower. It took me about thirty minutes to shower and dress—I simply didn't have any energy. I could barely lift my arms. I told my friends that I didn't think I was going to be able to go to dinner with them. They suggested I go back to their house for a rest, adding that they thought I looked a little pale, and that's what I did. I laid down on their couch and after another thirty to forty-five minutes of that, I still was overwhelmingly tired. Plus, I was getting chilled—I couldn't seem to get warm. At that point, they decided I should go to the hospital. The folks in the Emergency Room knew immediately I was having a heart attack and I was wheeled upstairs for stent surgery. I emerged a few days later "repaired."

If only the store could have been rebuilt that quickly.

The doctor who treated me told me that my cholesterol, blood pressure, and all other health-related tests came back in good order. He believed my heart attack was brought on by the stress of the fire and the insurance hassle. He was willing to testify to that end if a lawsuit was filed.

In the combination of insurance and health bills . . .

I had to go back to my bank and took out a new mortgage. That was depressing. I only had about four-and-a-half years left on my old note for the store—an amount of about $52,000. Without any input from me, the insurance company had paid off that mortgage note. The bank held a lien against the equipment in the store (a second mortgage) even though they knew they were to receive the insurance money. The bank called the second mortgage. It was a financial nightmare!

I scrambled my way to another bank for a construction loan and in the end, my attorney advised me to go court.

The issue ended up in the Wyoming Supreme Court. Wyoming doesn't have circuit courts, so my case went to the district court. The district court made a decision in my favor. The insurance company appealed the matter to the state supreme court. The judges of the supreme court also ruled in my favor, with the exception of a couple of items, and sent the matter back to the district court to resolve those items. One of the unresolved items was the amount of punitive damages related to my heart attack.

The insurance company had the option to refile with the state supreme court rather than leave it in the hands of the district court, with the understanding that the supreme court would decide on all matters related to the case and there would be no further appeal. This time, the supreme court ruled in favor of the insurance company. The decision was based on the fact that I had all of my finances set up in a living trust, and since a "trust" could not feel pain and suffering, the trust could not be paid punitive damages. *Good grief.* The living trust was only related to my personhood, and only kicked in upon my death. None of it made any sense to me—certainly no "common sense" was involved.

I needed additional legal counsel when the lawsuit first went to the supreme court. My adjustor advised that I consult one particular firm that had lots of experience in arguing before the supreme court judges. Although this firm did not take individual cases, they decided to take *my* case after they had reviewed it, certain of a settlement to me, which meant payment to them. They took my case on a contingency basis. When the final settlement came in, they were flabbergasted. They lost more than $100,000 in billable hours. I felt bad about that—but also bad for *me*.

The entire process lasted four years.

I must admit I was in awe when I went to the Wyoming Supreme Court. It was a process that seemed foreign to me according to what I knew about lower-court protocol. I can only imagine how a person might feel if his case went to the United States Supreme Court.

A person's attorney has a set amount of time to speak before the supreme court. The defendant himself, or herself, has no opportunity to speak. I recognized that my case wasn't being argued precisely according to the facts of the case. But I had no recourse. Nobody seemed truly interested in the *facts*.

A Little Time Out

As you might imagine, the entire four-year experience was exhausting. Somebody asked me one time if I was "burned out." Well . . . literally, I had been! The fire was a traumatic experience on a number of levels. Emotionally, however, I never felt truly "burned out" when it came to the Buford Trading Post as a business. I was emotionally wrung-out from the insurance hassle, and during those same years, the illness and death of my mother.

I had never been one for taking extended vacations from the store . . . now was the time.

Jonathan had moved out of state and I felt no constraining ties to remain in Buford year-round. So, in 2004, I leased out the store to one of my wholesale vendors who wanted to take on the challenge of running the Buford Trading Post. I had known this man for some time and I trusted him.

And . . . I took to the open road . . . on a motorcycle!

I was gone for as long as thirty-four days on one trip. Talk about freedom!

I took several extensive trips. In summary . . . I've been in every state, some of them multiple times. I've been on all four highways in Alaska, and I flew to Hawaii and rented a motorcycle there. I have also ridden the full length of lower Canada, including crossing the bridge over into Prince Edward Island.

A number of years ago I set a goal for myself of swimming in the Arctic Ocean. That's still on my bucket list. I was stopped about two hundred miles from the Arctic because the motorcycle I had rented was not considered "insurable" for the gravel road—frequently a muddy gravel road—that stretched out ahead of me toward the ocean.

My favorite rides have been in the mountains. I like the scenery and the temperature. I'm not eager to repeat rides across the deserts or central plains of the United States.

One of my favorite rides was the full two hundred miles of the Blue Ridge Highway.

I also like the ride down to Key West. I've probably made that trip fifteen times. Driving the Keys has all the things I like—

bridge after bridge, all in tropical splendor. I also like the ride on I-10 across Louisiana. Again, it's all about bridges. I'm not sure why, but I really like crossing bridges on a motorcycle—the higher and longer the bridge, the better I like it.

The trips by motorcycle gave me a slightly new perspective on the American traveler. I had been dealing with I-80 travelers for decades, but suddenly I *was* a "tourist."

I came to several conclusions:

• The best way to get in touch with America, and Americans, is "on the ground." Flying over America gives a person a distorted view—especially if you are going coast to coast.

• Americans are a lot more traditional and conservative than many of the media pundits seem to think. We are a generous people, very kind to strangers (for the most part), and we take our home-related responsibilities seriously. We like our freedom—especially freedom to make our own decisions related to our own earnings, jobs, associations, and freedoms of speech and self-defense.

• There's a lot of mobility in our nation—people moving, people being transferred from job to job—but much of that mobility seems focused on our cities. People in small towns and in rural areas are a lot more "stationary." They tend to put down roots and ride out the difficulties of life.

• Americans seem to take the major issues of life very seriously but they also are a fun-loving people! Nearly every place I've ever visited has its own festivals and parades and form of "fun" that is unique to that town or region. We like to laugh.

Are these the traits I saw in the people who entered the

Buford Trading Post?

For the most part, yes.

I liked the folks I met on the road.

I liked the folks that came in the door of my store.

And the good news is that both groups of people seemed to like me—at least enough for me to buy or sell them the "stuff" they needed, to eat or provide café food, and to give directions and information about local sights.

My vendor friend gave up the lease in 2006 and by that time, I was ready to take on the challenge again. I felt personally refueled to take on some new ideas and set some new goals. I actually let him out of the lease a few months early so I could stock up and gear up for the summer season, and I jumped back into the business with good energy and vision. I had spent considerable time during the leased-out years evaluating ways in which my new store might be different from the old store. It was a time of mentally sifting what should stay the same, and what might be changed or eliminated.

I was ready to go back into retail!

14

TOWN CREATION

GOING BEYOND A QUICK-STOP STORE

I had an intuitive sense for many years that it was up to a seller to SELL. We have so many choices of products and services in our nation. Our marketplaces are a buyer's paradise. What makes a customer purchase *your* product or service over that of another seller? Those are the reasons that a seller must isolate and push.

Most people gravitate toward factors that seem to provide something extra, or some form of added value.

I had a personal experience with this decades ago when I was in the moving business. I nearly always took with me a small motorcycle in my moving van. I loaded it just as I did other supplies such as tarps and ropes in the back of the rig. After I had unloaded a haul, and especially if evening was approaching, I'd use that motorcycle as my transportation to a nice restaurant to reward myself for a long day's work with a good meal.

One time I was in a restaurant and noticed that the menu featured a steak smothered in fresh mushrooms. That sounded good to me and I placed my order.

When the steak arrived, the topping was obviously one made with canned mushrooms. I'd been cooking for enough years to tell the difference immediately. I said to the waiter, "What happened to the fresh mushrooms?"

He replied, "These are fresh mushrooms."

"Aw come on," I said. "I know the difference between canned and fresh. These are obviously canned mushrooms and I want what the menu stated."

He left my table and returned a few minutes later with a can in his hand! He pointed proudly to the label that said "Made from fresh mushrooms"!

I ate the steak and chalked up the experience to a learning experience—for me, obviously not for the waiter.

People want to receive something "special." They want to have their imaginations tickled, their senses triggered, their sights elevated. Most people are intrigued by something unusual or unexpected, or something that triggers their curiosity. They'd rather purchase or pursue *that* than something that is ordinary, run-of-the-mill, or average.

At the same time . . .

People want what is advertised and they are disappointed when what is advertised doesn't meet their expectations.

It's a fine line to walk, but a line worth exploring and establishing!

I knew that travelers along I-80 would be making periodic stops for fuel if they were on a long-distance trip. They would need to stop *somewhere* for beverages, snacks, or food—or to use a restroom. My goal as a SELLER was to make sure they stopped at Buford.

Every vehicle that drove past the store was a customer

who hadn't stopped, and therefore, hadn't had the opportunity to buy something from me. Getting a customer to *stop* was the primary goal.

I heard one time that people buy the "sizzle" of a steak as much and sometimes more than the steak itself. I certainly knew that to be true in my life when it came to fajitas at a Mexican restaurant.

So, how does a person sell "sizzle" when it comes to a quick-stop gas station and store along a major freeway?

I'm not exactly sure when the idea of positioning the Buford Trading Post as a "town" first began.

It may have been the accumulative attention I paid to city limits signs with their populations posted along any number of roads that I had traveled over the years. It may have been the growing conviction that everybody referred to the store as Buford, as if a place named Buford existed apart from the Buford Trading Post. It may have been a growing recognition that I had made significant additions to the old Lone Tree Junction Store property, adding buildings and services—creating much more than a gas station and store.

Whatever the source of the idea, once it came to my mind, I ran with it. I saw it as "sizzle."

A second idea quickly came to mind, "We should have a sign!" Mail was coming to Buford, so I figured it was a town, and a town with a name. But there was no sign on the highway indicating that. I ran with that idea, too, all the way to a county commissioners meeting in Albany, Wyoming, and there, I requested a sign.

I wanted it to say "Buford, Population Don and Jonathan" and add the elevation. The commissioners informed me that they didn't have a problem giving me a sign, but they did require that the population be a *number*, not names. Initially we were "population 2." The commissioners asked me how I knew there were only two residents. I said simply, because I know both of them. I live there and the only other person who lives there is my son.

And, we added the elevation: 8,000 feet.

Buford *does* have the highest elevation along I-80, US highway 30. The actual summit is twelve miles away but there's no community there.

There was never any survey to determine the "city limits" of Buford. The sign was put at the edge of the highway at the two-way road that led into the store parking lot, on the spot where the state property butted up against my property.

After my son moved away, I notified the county and soon thereafter, a new sign stating the revised population arrived: Buford, Population 1.

The Smallest Town in America

It was at *that* point that I began to call Buford the "smallest town in America."

I sought a little legal counsel about that. It's one thing to call yourself the nation's smallest town—another thing to put it in print for distribution beyond the driveway to your store.

The attorney advised me that the definition of "smallest" was one I could probably defend. The exact acreage didn't really matter. *Small* could be a matter of population, and there could be no town *smaller* than one resident. If another town with one resident showed up, I might have had to negotiate, but until then, the entire matter seemed to be one of "no harm, no foul."

I also got advice about using the word "town." Buford had been a place that thousands of people had called home for more than a century. But the trading post store that sprung up there had never been incorporated as a town. Nevertheless, it did have that ZIP code: 82052!

The United States Postal Service had decided decades ago that direct delivery of mail to the widely scattered addresses in the region was too time-consuming. They also took pity on the residents in the region to the degree that they did not believe it necessary for them to drive into either Cheyenne or Laramie—the cities in the general area—to get their mail from a post office box there. Buford was about equidistant between Cheyenne and Laramie, so a bank of post office boxes was installed there. For years, I served as the postmaster, sorting the mail into boxes that were rented out, distributing stamps and sending packages, and so forth inside the store.

In retrospect, I also believe I could have defended the use of the word "town" on other grounds—Buford had most of those amenities and services and buildings that make a town.

It had a residential section—my house and a separate garage for vehicles.

It had a business section—primarily the store, but also a tow-truck business, a salvage yard, and a set of electrical poles for use by those who wanted power for their trailers.

Within the store, there was a café serving breakfast and lunch. And, I was very happy to cook a frozen pizza for those who had trailers parked on the grassy stretch nearby.

The store sold gasoline and diesel, various sundry items, and an interesting variety of souvenir items.

The tow trucks had a snowplow capability.

The property also had a bank of post office boxes, as well as a cell tower—so it could be considered to be a town with communications capabilities.

The parking lot was used by Federal Express nightly for a transfer of trailers from one tractor to another.

While the town of Buford had no municipal utilities, I certainly kept a clean restroom with an outside door, and kept the inside of the store and the surrounding grounds neat and orderly. I collected sales tax and passed on revenues to the state, including fuel taxes.

In all, it was a compact town of 9.9 acres, a good business venture, and a good neighborhood with no rival cities and virtually no crime rate!

And Guess Who Became the "Mayor"?

It was like icing on a cake that the Wyoming state transportation department granted me a sign. And everything felt very official the day a couple of guys in a state-owned vehicle showed up with a green highway sign and asked, "Where do you want *your* sign?" I showed them. They installed it. And so it was—photos of that sign were eventually published around the world.

With a proper highway sign, a ZIP code, and eventually a

population of only one after my son moved on to embrace his own life . . . one of my employees dubbed me the "Mayor of Buford."

Please note that I didn't call myself the "mayor" of Buford. Periodically, a customer would come in, quickly determine that I was the only person in the store, and when they found I lived on the property, they'd quip, "Well, then you must be the mayor, the chief of police, the sheriff, and every other city official."

I'd laugh and say, "Yes, and I'm the plumber, the janitor, and the parking lot sweeper."

They'd joke, "You must make all the rules."

"Yes," I'd agree.

And then I'd add, "And I have to keep the rules I make."

Sometimes locals would be in the store when a traveler would come in, and ask, "Is this the guy who *is* 'population 1'?"

They'd say, "Yes. He's the mayor!"

Please . . . just call me Don.

The Best Promotional Idea I Ever Had

I began to promote the Trading Post store and gas station as "The Nation's Smallest Town." I registered the name Buford Trading Post, and I enjoyed putting the name on lots of different items, even twig pens (ballpoint pens made out of real twigs from real trees). It wasn't long before national and faraway publications picked up that bit of news and began to write about Buford.

I wasn't into the idea of "fame" as much as I was into the idea of *advertising.* Good advertising meant that more people stopped by the store as they traveled on I-80, and the more vehicles that stopped, the more sales increase. Basic economics 101. I wasn't in business to gain personal notoriety. I was in business to sell stuff.

Nevertheless, the nickname "nation's smallest town" *did* make Buford famous.

And various national media outlets helped propel that fame.

February 7, 2011, was an exciting day for me. I don't particularly remember that *date,* but I do have evidence of what happened on that date. The *National Enquirer* had a feature article about Buford with a photo of me standing in front of the road sign that declared the population of Buford to be "1" and the elevation to be 8,000 feet. It was great publicity for the Buford Trading Post.

That article was a bit of validation for me that I had been wise to get that sign!

I discovered increasingly that people were intrigued with the *idea* of Buford—of one person building a property that might be called a town, in the middle of nowhere . . . and liking it.

What became a major selling idea for *product* in the store eventually became the major selling idea for the store itself.

But first, let me digress just a bit . . .

15

INDIVIDUAL STYLE

DOING THINGS MY WAY

You may have concluded by now that I'm a guy who likes to do things "my way." I think of it as creativity, perhaps rugged individualism, perhaps old-fashioned ego, perhaps the quest for challenge and adventure.

Some time ago, a person said to me, "I can't quite figure you out. Are you a daredevil who lucks out, or a guy who is just quirky and fun? Or are you one of the best marketers and promoters in North America?"

I thought for a moment and then replied, "Do I have to choose?"

I've never been one for making long lists and analyzing data. My choices in life and about the Buford Trading Post have been far more spontaneous than data-driven (except, of course, for which items to keep on the shelves of the store). I'm a guy with strong personal likes and dislikes, and a guy who likes to try new things.

Lest you think that I've only been a store owner who shoots out tires and takes long motorcycle trips, let me share a couple of other things about Don Sammons, the Mayor of Buford.

A Love Affair with Mountainscapes

The mountains have always filled me with awe.

Mountain scenery does something to me that other types of scenery just don't. Although I've lived close to the ocean for a number of years of my life, I was never really inspired by the ocean. A beautiful sunset could be inspiring, of course, but even then, I was more focused on the sun sinking into the Pacific and in the colors of the sky, than the ocean itself.

The flat prairie of the heartland was like a dry ocean to me.

Yes, give me the mountains!

There's so much variety in the visual effects—a scenic buffet of peaks and valleys, evergreens and prairies, snow and glaciers, waterfalls, deer and elk and eagles.

I have a longstanding friend who had a ranch in western Wyoming and later moved to Cheyenne. She is planning to move to Sedona, Arizona, and she tells me how beautiful she thinks the red sandstone and desert environment is there. I've seen the pictures. I don't see what she sees.

I have never become tired of the mountains and general scenery of Wyoming. Even if I have been down a highway a hundred times, there are moments when the sheer grandeur and size of the Rockies takes my breath away and I find myself whispering, "Wow." I suppose others feel that same sense of awe at the beauty of nature when they are in the desert or along the ocean, but neither desert nor ocean causes my heart to skip a beat. Snowcapped peaks . . . dense green foliage . . . crevices and canyons . . . crags and waterfalls . . . now *that's* natural beauty! It has been that way from the first day I drove through the Rockies. It is still that way today.

If it's a good idea to live in the general vicinity of the scenery that inspires you, then my move to Wyoming was a GREAT idea.

After the shootout incident, the court ordered me to see a psychologist in an attempt to determine if I was a danger to the public. She was about seventy years old, with many years of experience in counseling and analyzing. She concluded that the reason I had moved to Wyoming was *because* of my Vietnam experience, far more so than my weariness from the southern California hassle and traffic. From her perspective, Cheyenne was the "high ground" to me, far from an angry population, far from war, and with such wide-open spaces that virtually nobody could hide in the foliage there. I'm not sure about all that. What I knew was that I was drawn to the mountains—their beauty, their grandeur, their permanence. I didn't need any other kind of high than the "Rocky Mountain High" I got from looking at the scenery.

Soaring above the Wyoming Skies

I've enjoyed looking at the mountains from as many angles as possible, including "down" from an aircraft.

I once took lessons to become a pilot, and the lessons involved a Cessna 172 single-engine plane. The Cessna 172 is a marvel—well balanced, easy to fly, reliable, and noisy.

One of the techniques my flight instructor used was a tactic called "putting the pilot under the hood." I was required to wear a hat with a big brim—the idea was to force a new pilot to look at the instruments before him, not look at the scenery out the window. I was flying along one day and thought I was maintaining an even altitude . . . apparently I couldn't take my eyes off the scenery outside the plane's window. Suddenly I heard the flight instructor say, "I've got the aircraft."

That is not something a student pilot wants to hear. I was descending but I didn't "feel" as if I was descending. His words certainly were a wake-up call! They meant that I was *not* maintaining an even altitude. We were going down and he had waited for a little while to see if I came to my senses. When I didn't, he took over, and we lived to fly another day. I let go and let him have the aircraft! Permanently.

I didn't complete the Cessna training, but it wasn't because I hadn't learned a good lesson about watching the gauges.

I switched to *gliders*, primarily because the experience felt more like flying and because it was "silent." There was no roaring of an engine or wind noise, and in its place, there was a very keen sense of air currents. On several occasions, I looked out a side window of the glider and saw an eagle alongside me. Eagles don't fly with planes that have propellers and engines!

In a glider I wasn't bound to instruments. I wasn't surrounded with noise. And the entire experience felt much more like real *flying* to me. I loved finding the thermals and floating wherever they took me.

Flying a glider is quiet. A glider can be launched from a trailer that is towed by a pickup truck. A glider can allow you to soar just a few yards away from an eagle, who amazingly isn't spooked by a quiet "bigger bird."

Some gliders do have engines, of course. But they are not the mainstay of the flying experience. Find the thermals and float—that's the goal. It is a thrilling experience.

Most people don't realize that gliders can catch currents that will take them *extremely* high into the atmosphere. I knew a man who had contacted a plane on the "glider frequency" who told him he was at 75,000 feet. Another pilot was apparently in that upper atmosphere with him—even *higher*.

A tow vehicle will pull a glider down a runway with a cable, and once the glider is at a predetermined altitude—a couple of hundred feet in the air—the cable breaks away and the plane and pilot are free. If the plane needs to land, it can be turned to land on the same runway from which it was launched. To add to the accessibility, a tow vehicle might be a plane, or it might be a truck. It's usually a lot easier and faster to head into the skies if you're the pilot of a glider!

Tattoos and a Diamond Stud

Both of my arms are covered with tattoos. They commemorate my entrance into the U.S. military, and my honorable discharge from the military. I am proud that I served my country, and that I served in Vietnam.

The earring is also related to the military. I stopped at the Vietnam Memorial in Washington, D.C., on one of my extended motorcycle trips. I had thought it might be a fifteen-minute stop. I was there for hours. The monument captured my imagination and evoked a flood of memories that I didn't even know I had. I felt a great gratitude for those who fought and died in that war. They had been my comrades. And since there was no room for more tattoos, I decided to wear a large diamond stud in one ear. That way, when people might ask, "What's with the earring?" I could tell them about Vietnam and remind them that the men and women who served there were honorable patriots who believed in freedom, and were willing to die for it.

Driving Jaguars— As Fast as I Can

In my opinion, there are cars in which you feel you have to rush. There are other cars that give you a rush.

I have owned some very pretty, classy cars, but they just didn't excite me. I found them pretty boring, and once I started seeing the same make and model in the same color, I found them "average" even though they were luxury cars far beyond average.

I felt differently after I purchased my first Jaguar, a four-door sedan. A trip to the store in my Jag became an *event*. I wasn't just driving a car. I was having an experience.

Several years ago I purchased a Jaguar XK (sports model) and shortly after that, I took advantage of a special program that Jaguar has for its owners. In an effort to display the many engineering feats built into a Jaguar, the company offers select owners an opportunity to drive a Jaguar off-road in conditions that can really put the vehicle to a serious test when it comes to handling, speed, acceleration, and so forth. Most of these features of the car aren't known by an average driver who is limited to the constraints associated with city streets and highways.

I chose the Nevada "course" to drive a Jaguar and went to

the Las Vegas International Speedway, where an obstacle course had been set up, along with a timed event and an oval-speed-track event. My instructor for the experience was Adam Andretti, nephew of Mario Andretti. He taught me how to drive like a professional race-car driver and how to put the Jaguar to the "test" in various parameters.

The computer on Jaguars has a number of safety features, all of which *can* be turned off. When these features are turned off, you truly discover what the computer normally does to ensure safety.

It was a full day of racing and I got a real kick out of driving as fast as I could. Jaguar paid for the tires and for the gasoline. All I had to do was burn up the fuel and lay rubber. I had the opportunity to drive virtually all of the Jaguar models, and was pleased to discover that the four-door sedan was as much fun to drive as the two-door, souped-up sports car.

We did a time lap in the morning and then again, at the end of the day. By the end of the day, I had lowered my time by five seconds and came in third out of twenty-eight drivers at the track that day.

I learned a lot about straightening out curves, which helped me greatly in driving the mountain roads of Wyoming.

I like speed, of course. I considered for many years that the twenty-five mile stretch between Cheyenne and Buford was something of a personal racetrack. The fastest I ever covered those miles was twelve-and-a-half minutes. I'll leave you to do the math.

"Weren't you afraid you'd get stopped by the highway

patrol?" someone once asked me.

Not really.

In the first place, a Jaguar not only has cruise control but a "speed limiter." That's a feature few people know about. Cruise control allows a driver to just "aim" a car. The speed limiter is set by the car's driver, and then the driver can drive the car and the car doesn't say anything about the speed until the "limit" is met. Once I set the speed limiter I don't worry about getting a ticket because the car will tell me about that possibility in advance.

In the second place, the Wyoming Highway Patrol is not on a twenty-four-hour clock. The patrolmen are only called out after hours in cases of accidents or emergencies. After dark, the roads are . . . well, race worthy!

I wasn't worried about the highway patrol as much as I was concerned that an antelope might skip out into the road at the wrong time.

That never happened, and I'm grateful.

In taking my Jaguar on road trips to California, Louisiana, and the Florida Keys, I've never seen another one like *my* car. I like things that are unique. What else would you expect from a man who enjoyed several decades in a town of "one"! My choice of vehicles is a way of expressing uniqueness.

A Patron of the Cheyenne Symphony

A fact about me that few would ever suspect: I am a patron of the Cheyenne Symphony and a lover of classical music.

My father loved the guitar, and learned how to play it. Actually, he taught himself. He bought a guitar and played around on it until he could pick out some tunes. I thought I'd learn it, too, so we could perhaps play together. It didn't happen. It turned out to be a "neat idea" for only a couple of days. My father enjoyed country-western music, but it was never really my style.

As a young teen, I also tried to learn to play the saxophone, because I loved the sound of the saxophone, and I *thought* I wanted to be a sax player . . . but I hadn't counted on the idea of *practicing*. I don't think I had the discipline necessary to learn an instrument.

I also took a turn at drums. I could really beat those drums, but not with any sense of style or rhythm for being part of a combo. I rightly concluded that a musical instrument was not something that could be conquered in a day, and therefore, it was probably not something I would ever succeed at in my life.

I didn't have an ear for music. I skipped music class in school, and never developed an appreciation for different styles of music. I didn't know how to read music and wasn't much into singing. The same thing held for art—I also skipped art class in school, and to this day, have trouble drawing a stick figure for a person. I have opinions about what I like and don't like as "art," but I would never classify myself as artistic, or musical.

It was a big surprise that I fell in love with classical music at my first live concert.

It happened by accident.

A friend had season tickets to the Cheyenne Symphony and he offered me a ticket to a concert he couldn't attend. I recognized the ticket as a gift, and frankly, I didn't know how to turn down this gift without feeling as if I was rude or ungrateful.

Then, I couldn't think of any legitimate reason *not* to go, and I had no clue what I might say to my friend later about why I hadn't gone to the concert . . . so the long and short of it is that I went to the concert.

The visiting artist for the symphony the night I went was a woman violinist—probably in her early twenties at the time. I had never heard anyone play an instrument the way she played. I had never heard music that stirred something so deep in my soul. It literally was one of the most profound experiences of my entire life.

The next year, I purchased my own season ticket!

I still make no claim to understand orchestra music, nor to be capable of identifying different styles of music or even to readily recognize the sounds made by the different instruments. But I do make a claim to love classical music, and especially music that is played live in a concert hall. There is something about the experience that moves me and stirs a sense of life and beauty deep within me. It is a different experience than staring at the grandeur of the Rocky Mountains, but it is a *similar* experience in making me feel something far beyond and much bigger than myself.

The second year that I purchased a season ticket, I also contributed to the orchestra fund, and thus, I became a "patron of the arts" in the Wild West! Who would have thought it possible?

Today, I am a primary or contributing sponsor to one or more concerts a year. I consider it a great privilege to do this, and even though I have moved away, I return to Cheyenne for these concerts, and to a great extent, plan my schedule around them!

For years, I have also been one of the volunteers who either picks up or delivers the conductor of the Cheyenne Symphony to the airport in Denver. The conductor of the Cheyenne Symphony is also the conductor/artistic director for the symphony in Dubuque, Iowa, which is where he and his family live year-round. In the beginning of my driving for the conductor, I got to know Stephen Alltop, the previous conductor, and now I enjoy the company of William Intriligator.

The four-hour "conductor pickup" trips between Cheyenne and Denver have always been very enjoyable to me. I not only have had opportunity to get better acquainted with the conductor as a person, but have also gained insights into why he has chosen certain musical selections and certain solo performers.

I especially enjoy the classics, and of the classics, Mozart and Beethoven. Of the more modern composers, I like Copland.

I'm not particularly interested in knowing how a concerto is put together. I am extremely interested in watching the musicians perform—especially the solo pianists—and in experiencing how the music makes me *feel*. It transports me to a world I don't otherwise know.

Giving to the Cheyenne Symphony is a way of giving back. I have always believed that "giving" is a good thing to do—to as many as possible, in ways that are meaningful.

In years past, I enjoyed helping the little school down the road—giving donations occasionally that allowed the teachers to take the students down to art shows and museums in Denver. In gratitude, the parents of the students often referred business to me at the store. That was pretty much the spirit of Wyoming at work.

People huddled together in the middle of nowhere tend to take care of one another if they can.

16

SELLING BUFORD

WITH HELP FROM THE MEDIA AND THE PROFESSIONALS

The reasons I moved to Buford initially were the same reasons that kept me there for twenty-six years.

Basically, Buford offered a slower pace of life than what I had known in southern California. The slow pace may not have turned out *exactly* as I had envisioned it once I gave up the long-haul moving business and bought the store. I suspect, however, that running the Buford Trading Post has been a more relaxed business than running a similar place in Manhattan or Los Angeles. No doubt about one thing—once I had a house about two hundred yards from the store, I had no commuting problems!

The Buford area gave us plenty of room for horses, dogs, and cats. And, I did enjoy the solitude and the freedom to make my own choices. I certainly didn't have any problems with noisy or obnoxious neighbors.

I loved living in a log home on a ranch; during the years I did that.

I loved the beauty I saw all around me.

I felt as if I was living in the midst of an adventure of my own making. That was exhilarating to me.

Up until the last three years of my being in Buford, I truly *enjoyed* everything about the experience of living there and running the Buford Trading Post.

And then, I simply knew it was time to move on.

Gearing Up to Sell Out

For about a year before the auction, I had thought seriously about selling the store. I had contacted a couple of real estate agents to get their evaluations of the property and the potential price. After talking with them, I determined that a traditional sale wasn't going to be as fast, or yield as much money, as I had hoped.

I began to think, *Is there another way to do this?*

I remembered that a truck stop a number of miles east of me had been in an auction. I had actually attended that auction, as well as auctions of a restaurant and some other properties. I was familiar with the auction concept and process, but I couldn't find anybody who had ever sold a "town" like mine by means of an auction.

I also did some research into eBay. I was intrigued with that possibility, but just couldn't get comfortable with running the auction process by myself. I began to search for someone who might help me do an eBay auction, but I couldn't find anybody at that time. There were lots of people who knew how to buy on eBay, fewer but still lots of people who knew how to sell small items on eBay, but nobody who seemed to know how to sell an entire town on eBay!

I also knew that I would be wise to find somebody, or a

firm of some type, who had the ability to capitalize on the uniqueness of the Buford property. AND, perhaps most importantly of all, I knew that I needed a company that could cast a large net for buyers. I pretty much knew the business people of Cheyenne and Laramie, and I was convinced the buyer was not likely to come from either city. In my thinking, the buyer was going to come from another state, perhaps another nation. I needed to be working with realtors who could attract an international audience to the front door of the Buford Trading Post.

I talked to a man who had purchased a property at an auction and he was very happy with the company that had handled the sale. I contacted that company: Williams & Williams in Oklahoma. A few weeks later, the company sent a member of their team to look at Buford. I was looking to him for "particulars"—not just anticipated price and the necessary safeguards, but particulars about how Williams & Williams might *market* the sale to a very big audience. I wanted people who could see value in the IDEA of Buford, Wyoming, as a *town*, not a mere quick-stop store along a freeway. Intuitively, I knew that the IDEA of Buford was the most important factor in the sale.

I had done enough research to know that at any given time, there were probably fifty or more "convenience stores" for sale in high-traffic areas like I-80. What would make a person want to buy *my* store? I strongly believed that I was *not* selling a convenience store. I was selling a *town*. And the question then becomes, *What would make a person want to buy a TOWN?*

It's a lot more interesting to buy a TOWN than a store that

sells soft drinks and gasoline.

Initially, the guy from Williams & Williams couldn't seem to grasp the idea of selling a *town*. They had never sold anything on the basis of an IDEA being the overarching, main focus. This man only saw buildings and equipment and acreage. But once he went back to the home office, and he and his colleagues started tossing around ideas, they got excited about the idea of selling a *town*. And their excitement grew as the months went by.

I wanted the auction to be held by the first week of April. That way, if my minimum asking price wasn't reached, I still had time to gear up for the busy season that kicked in on Memorial Day.

In the meantime, I sold down my inventory. By the time set for the auction, I had sold most of the candy, snacks, soda, beer, and all the Marlboro cigarettes. Bags of charcoal, whistles made from animal antlers, and dozens of T-shirts remained unsold.

I wrote on my website, "This entire, income-producing town is for sale; the house, the Trading Post, the former schoolhouse, along with all the history of this very unique place."

Can you sell history? No, not really. But you can pass on the responsibility and privilege of being the "keeper" of the historical account of a place, and that's what I fully hoped and intended.

The chief marketing officer for Williams & Williams noted, "We're going to have a variety of people attracted to this property, based on what it would mean to them." *What it would mean to them.* That was the essence of the sale!

A Sound Business Investment

In selling the idea of Buford, none of us ever lost sight that it was a sound business investment for somebody with marketing savvy who was willing to work.

My minimum asking price was $750,000.

The starting price for the auction was set at $50,000. I thought that was too low. The company, however, countered that more people would come if the auction began at that level, and the more people present for the auction, the more excitement would be generated and the higher the price would go. All I could envision was a group of people who walked away disappointed, angry, or otherwise disgruntled by standing in the Buford wind only to have the auction called off because my minimum selling price wasn't reached. I also figured that a serious buyer might not be interested if the starting price was low—that person might not think the business was viable and wouldn't show.

The real-estate folks and I finally compromised with a starting bid of $100,000.

Several people have asked me since the auction, "How did you set the minimum sale price?"

It was entirely a financial calculation.

I knew how much money *could* be made at the store! For a number of years, I grossed more than a million dollars annually in sales. That certainly wasn't the *net* after all the costs related to fuel, goods, employees, taxes, maintenance, and so forth were factored into the equation. But, if a person didn't go crazy on purchasing, a good living could be made!

I also knew that for the previous decade or so, I had about 100,000 customers between Memorial Day and Labor Day. That's more than a thousand people a day. The majority of the customers purchased fuel. Some were repeat customers. Even out-of-state travelers often became repeat customers—stopping on their way east, and then again on their way west, or vice versa.

Not only was there a solid base of customers through the front door, but there were the other income streams I detailed earlier:

- The cell tower generated a small income.
- The electrical posts generated income.
- The bank of post office boxes generated a small income.
- The 2 AM Federal Express "exchange of trailers" generated a small income.

Add all the "small income" streams together and it was a sizable amount . . . plus the income from the other nine months of the year (mostly fuel purchases and lots of coffee).

From the outset, I planned to sell the tow trucks and snowplow separately. By the time of the sale, I had three tow trucks. I knew the towing and salvage yard businesses were good income producers, but at unpredictable levels.

In all, the "town" was profitable!

Plus . . .

The three-bedroom house and all its furnishings—from teacups to towels, from furniture to appliances—was part of the sale. The home was fully stocked, right down to extra garbage bags and paper towels. The owner only needs to turn on the thermostat to the setting of his or her desire. The sheets and towels were clean and ready for use.

I Began My Personal Transition Elsewhere

I bought a house in Windsor, Colorado, several months before I sold Buford. I took personal items and a couple pieces of furniture that had belonged to my mother or grandmother, but, for the most part, I started over in purchasing all the basics. New things for a new place as part of a new start!

At first, it was fun. New dishes and silverware, new furniture and bedding, new electronics and appliances. But after a while, I thought, *Don, what WERE you thinking?* I'd suddenly realize that I didn't have a can opener, a strainer, a toilet bowl brush, a soap dish for the kitchen sink, or a towel for the guest bath. I seemed to be making endless last-minute runs to the local hardware store, or home-furnishings outlet. I was *very* glad when that came to an end.

As a former long-haul mover, I knew that some stuff is worth paying to have it shipped, and other stuff can be purchased at the new location for less than the freight charges. In my case, however, it wasn't only a matter of money—it was a matter of "starting fresh."

I moved to Windsor in the fall before I closed the store. I commuted back and forth every week. And then after I closed the store the first of the year—the auction not happening until April—I drove up to Buford about once a week just to check on things.

This was important for me as a transition. It gave me time to sort out what I wanted to move and didn't want to move. It gave me time to find a "landing place" after Buford.

An Escalation of Media Interest

That old saying about a "prophet is without honor in his home town" was one I found to be true when it came to publicity from the local media outlets in Cheyenne and Laramie. In all the years prior to the sale, nobody at the local newspapers or television stations saw the uniqueness of Buford as something newsworthy. Perhaps that was because an independent spirit is so prevalent throughout all of Wyoming. Perhaps it was just a lack of a keen perception. Several people have sought to capitalize on the fact that if they had MY support for something, they had the support of an entire town.

That isn't to say that media interest was missing. I had a number of very nice encounters with the media, and those seeking to become famous, through the years.

Mark Gordon was a rancher in the area who ran for the U.S. House of Representatives and he made a concerted effort to visit as many businesses and residential areas as he could. I listened to his thoughts one day and I told him that I would be willing to vote for him and back him. He asked me if I would participate in a promotional spot for his campaign. I agreed.

He sent out his PR people and they produced a thirty-second spot that featured Buford. They had flags flying from all of the fence posts leading from the highway to the store. Gordon sent the message that he wanted to represent EVERY voter in Wyoming, including the guy who lived in a town that had only one resident. One thing for sure—he had one hundred percent support from the voters (er . . . voter) in Buford!

Alas, he didn't win.

I also had several people come to Buford to tape a segment for *their* television programs. The Buford Trading Post was a unique "backdrop" for them.

Perhaps my most interesting media encounter was one that didn't really "happen."

In 2010, I got a call from Maria, "the Korean bride." She wanted to know if I wanted to marry her. This gal was a television producer in New York City who did documentaries about different "kinds" of lifestyles in the United States. She would find a man with an interesting nationality, career, or location, and then create a plot line focused on the question, "What would it be like to marry this person?"

Maria Yoon's documentaries aired in South Korea, and part of her purpose—at least from my perspective—was to show Korean women that it was acceptable to marry somebody other than a Korean man and to live a life that was not conventional by Korean standards. Each documentary was staged as if it was a courtship between Maria and the man with the unusual story, and each documentary culminated in a mock wedding—not legal, but certainly interesting!

Her goal was to "marry" somebody from every state in the lower forty-eight states. I was her target for Wyoming.

She told me later that I was the only person who ever

turned down her marriage proposal.

I thought it was a scam, and only later learned that it might have been a good money-making opportunity for me. She went to Laramie and found a guy there who would marry her. After the "wedding" to this guy, she drove up to Buford to meet me—she was intrigued, I think, by the fact that I had declined her invitation.

We became friends and are still in touch by e-mail. No, there's no wedding in the future.

After her documentary wedding in Wyoming, the Korean national television network did a documentary on *her* and the various locations she had chosen for weddings. These TV folks ended up in Buford for part of their story about Maria. They were intrigued by the story of Buford and in the end, there was good publicity for *me* apart from my becoming a "husband."

After Williams & Williams and I finalized our agreement, their advertising and promotions company went into high gear to contact radio and television outlets, as well as Internet and print outlets. I think the company began to see not only the potential for an actual high-dollar sale, but also that there was public relations value for Williams & Williams related to their handling the sale of a town.

The realtor issued a packet for potential buyers to prequalify those who might be interested in bidding—which included profit-and-loss statements, surveys of the property, and so forth. Requests for the packet came in, along with requests for interviews, from *The New York Times* to overseas television networks.

Over the course of several weeks, the Williams & Williams webpage had more than 36,000 hits, and 500 media outlets contacted the company for interviews. Inquiries about the sale came from 110 nations.

I personally did 238 separate interviews.

On most days during the six weeks before the sale, I was scheduled to do two or three major media interviews. Most of them were done by phone, especially those for radio stations (such as Canada radio). Some of those doing interviews called me a number of times to verify information or get updates. I felt as if I knew several reporters by the time the sale actually happened.

In addition, I had a couple hundred more e-mails commenting on the sale and the uniqueness of the Buford Trading Post property. The most interesting e-message was from a family in Costa Rica. They lived in a tree house!

One of the most interesting and unusual interviews was a taped interview for the equivalent of CNN in China. I couldn't really fathom why people in China might be interested in the sale of Buford. Frankly, I couldn't fathom why the news crew, based in Washington, D.C., would even want to leave the exciting news world of our nation's capital to travel to Buford, Wyoming. I asked the Chinese reporter doing the interview, "How many people do you think will watch this program you are taping here in Buford?" I was blown away by his answer, "About two billion." Two *billion?* The reporter informed me that their network was broadcast not only in China but other nations of Southeast Asia.

I found it exciting to be on the number-three news page of *The New York Times*. I was interviewed by both the Russian television network and the Korean national TV network for "specials" to be aired nationally. Calls came in from the BBC and from a radio station in Ireland.

In the cases of both the Russian and Chinese interviews, I recognized that the sale of an entire "town" was something that couldn't happen in either of those cultures. I asked the Chinese reporter why his network was interested in doing a story about Buford. He replied, "Because this could not happen in China!" First, nobody could *own* a town in China. Second, nobody could live *alone* as one person on ten acres of personal and commercial property. It was an entirely novel idea. The same was true for Russia.

In a way, I felt I was doing my part for international relations and capitalism in general.

The top three questions asked of me—whether from American or international reporters—were these:

How do you have a town?

How do you sell a town?

What are you going to do now?

I became skilled at changing my answers just a little so as not to get bored with my own story.

As the weeks went by, momentum seemed to build. I wasn't stressed or nervous about the sale. I enjoyed meeting the various people who showed interest in the town I had created. The experience, for the most part, was fun and exciting. I felt very good about the process, and the sale. I don't think it really hit me until the day of the sale that the "end" of my Buford days was approaching.

On the day of the auction, about thirty media outlets were represented on site.

Amy Bates of Williams & Williams noted, "Auctions always bring a lot of attention, but even we were amazed at the amount of attention to Buford worldwide. It's the Wild West in the U.S. It's owning your town and getting away from it all."

By April, we were ready to rock and roll.

17

THE AUCTION

EXCITEMENT AND CHANGE IN ELEVEN MINUTES

On the Thursday morning of the sale, I awoke early, checked my e-mail, ate breakfast, put on a black suit, and headed out for the seventy-five-mile drive to Buford.

I pulled into the Buford Trading Post parking lot just after 9:30 AM. My phone was already ringing. Representatives from the auction company pulled me aside to discuss the day. People inside the store came over to shake my hand and a couple of them told me they had seen me on television the night before. Still others had questions about the snowplow, the cell tower, and how to turn on the gas pumps.

Pam McKissick, co-owner and CEO of the auction house, advised me that potential bidders included a range of people from across the United States, Europe, and Asia, as well as a few wealthy and local Wyoming residents.

I heard Pam tell a reporter who had entered the store, "I think in this time when people are having trouble with gas prices, they can't pay their mortgage, maybe they're disgruntled about the economy or their jobs, this seems like a magical place where you can come to the great West and make a living and be in total control. I think there's a romance about that."

At ten o'clock an announcement went over loudspeakers outside for potential buyers to register and pick up a yellow bidding card. I had questions from a few reporters and requests for photos. A TV microphone was clipped to my lapel.

Reporters mostly wanted to know:

- *What is today like for you?* "Bittersweet. Buford's not going to be me anymore."
- *Is Buford really the smallest town?* "I'm the only one I know with a ZIP code."

By eleven o'clock, there were thirty-three cars, trucks, and vans in the parking lot. During the next thirty minutes, two more arrived and I did an interview live with CNN.

Just before noon—the set time for the auction to begin—a crowd gathered near the gas pumps.

Exactly at noon, the Williams & Williams auctioneer, Sonny Booth, called for bidders and onlookers to move closer. There were about fifty people gathered on the parking lot. Many of those were media people. Some were friends I recognized who were just curious to see what might happen.

The auctioneer asked if those gathered had ever heard the phrase, "I own this town." Then he added, "Well, ladies and gentlemen, *you* can own this town."

I took a photo on my phone. A friend patted me on the back. Booth continued, "Does anyone have any questions before we start the auction?" There were no replies. Then he said, "If not, we're going to start the bidding, ladies and gentlemen, at 100,000 dollars."

Ring men, some on phones, scanned the crowd, pointed,

jumped, and shouted as bidders began to raise their yellow cards.

"All right, I got 100, thank you. I got 100 here, I need 150..."

And so it began. The price climbed—200, 300, 500, 600, 700.

I was scanning the crowd, trying to see who bid, but the bidding was so fast I really lost track of the numbers.

At $850,000 Booth asked, "Anybody want to jump in?"

A few moments later, the gavel went down. At 12:11 PM, Buford sold for $900,000. The entire process had lasted only eleven minutes.

It turns out that only about a dozen people entered the bidding process, but that was enough. The opening bid of $100,000 had escalated upward very quickly. I wasn't even aware when the bidding passed my minimum asking price.

I also didn't hear when the gavel came down to indicate the sale was final, and I didn't hear the auctioneer say, "Sold!" I had been looking to see who had placed the latest bid. I was standing with the owner of Williams & Williams and it wasn't until she said, "Well, it's sold," that I realized the auction was over!

I fought back tears once the reality of the sale hit me. I had a moment of realization that my "baby" was gone—those were probably feelings similar to those who see their child drive away to college, or walk behind the closed doors of a kindergarten classroom. Buford was no longer mine alone.

I said to a reporter nearby, "I don't know when it will *really* hit me. I've lived here almost half my life. I'm an emotional person, and I hope I can handle this in an adult manner."

I knew my life had changed. I just didn't know how.

Meeting the New Owner

After the sale, there were papers to sign and windows to be boarded up.

After the sale, I walked into the store to sign the necessary paperwork and to meet the man who had placed the highest bid.

What a surprise that was—perhaps the word "surreal" is more appropriate.

Pham Dinh Nguyen and another Vietnamese man from Saigon were presented to me as the buyers! *Hardly the image a person tends to associate with Wyoming!* I'm not sure I had ever seen a person from Vietnam in Wyoming prior to that day—perhaps a Vietnamese person had come into the store at some point, but I don't recall it. Perhaps I saw a Vietnamese person in Cheyenne in a store or restaurant, but I don't recall it. And yet, there he stood, the new owner of my town. *From Saigon!* Now called, of course, Ho Chi Minh City.

The irony hit me immediately. I had spent a couple of years of my life in his nation—in conflict. And here he was, in peace, standing in my native land, buying my town.

Nguyen, age 38, initially wanted to remain anonymous but now that his name and more information has been made public by him, I feel free to share a few details.

The auction was a first for Nguyen, and so was his visit to Buford a first. He told a reporter that "waves of skin-cutting cold blew into my face," but then he added, "however, I was undeterred because of the desire to own this town." He also said, "Owning a piece of property in the United States has been my dream."

Back home in Vietnam, Nguyen apparently got a bit of notoriety and applause for "showing the world that Vietnam has moved far beyond war and poverty."

Nguyen runs a trade and distribution company in southern Ho Chi Minh City, and he initially told reporters there that he expects to use the Buford Trading Post to sell items made in Vietnam. He told Vietnam media, "Frankly, I just see Buford as part of the United States—a large and potential market for Vietnamese goods. Buford is likely to be the showroom for such goods."

A fellow businessman in Vietnam, Tran Thanh Tung, said to media gathered to cover the auction that he was "surprised, but also proud." He noted that the purchase was "something that one could not imagine a few years ago."

Not everybody in Vietnam was enthusiastic about his purchase. One person noted publicly that Nguyen would have been better off to spend his money in Vietnam helping the Vietnamese economy.

When it came to specific plans for Buford, Nguyen said, "To be honest, I do not have a specific plan for the town. But I think we Vietnamese should not feel inferior. Nothing is impossible." The fleeting thought occurred to me: *Perhaps the IDEA of owning Buford is the main benefit of his purchase.*

Two other factors about the buyer stand out to me as I reflected further on the sale.

First, I learned later that when Nguyen returned to Vietnam, his father greeted him and told him that he had seen the video story about Buford that aired on the Chinese news channel! Hurrah for the CNN of China!

Second, the buyer was a Buddhist. What are the odds that the only Buddhist in the greater Buford region would sell to a Buddhist? I suppose in Buddhist philosophy that wouldn't be an amazing fact, but it still struck me as highly unusual. I definitely had a sense of *oneness*.

As Reality Set In—A Little Celebrating

As minutes turned to hours after the sale was final, I felt a huge wave of relief. I had come to sell the store and the "town," and it had sold. I also had feelings that I had made the right choices—the right choice to sell, the right choice of real-estate company, the right choice in timing. I felt that Williams & Williams did everything they knew to do, and they did what they had said they'd do.

In looking back, I think everybody who came to the auction had a good time.

Several Buford-area residents were interviewed. They noted that they anticipated having "nice" new neighbors. They expected the community feel to continue at the store. As one nearby resident said, "The Trading Post is a place to get caught up with your neighbors, shoot the breeze, learn what's going on and who is around."

Some of the residents in the area have lived there for decades, some for their entire lives. One woman, Lucy

Williamson, told about her first visit to a store on the site of Buford—they had run a small Mexican restaurant.

Most of the longtime residents didn't see Buford as changing—they saw the sale as a new chapter in a long and familiar story.

I personally went out with a former employee of mine to enjoy a very fine dinner. We celebrated in Fort Collins. I spent most of the evening trying to let the reality of the day soak in.

I had an increasingly warm glow of satisfaction and a little nostalgia.

My work in turning a gas station and store into a town had been a creative endeavor. Letting go of a creative endeavor can be a little tough. I'm not sure how artists and sculptors feel when a project they've worked on for a long period of time suddenly sells and goes to live in somebody else's place. But I imagine that some of them feel a little the way I did.

I had succeeded in a major goal when it came to marketing. I *literally* had made Buford, Wyoming, and the Buford Trading Post, world famous.

And By the Way . . . The Check Cashed

There is a lag between the end of an auction and the time a purchaser's money actually hits the bank—in other words, a period of two to three weeks for final details of the sale to be wrapped up and for the check to cash. I wasn't concerned that this buyer didn't have the necessary funds. He had posted the $100,000 amount required to participate in the auction—which was applied to the sale.

Still, I was relieved when I knew that the money had been wired to Williams & Williams, and then to *my* account. That fact had nothing to do with his being Vietnamese—it had everything to do with my feelings of finality about the sale.

BUFORD had SOLD.

18

NEW HORIZONS

AND TIME TO REFLECT ON LESSONS LEARNED

It took me several months after I sold the store to regroup. I told several people that I was trying to get my ducks back in a row—the sale and the move seemed to scatter my "ducks." I felt that lots of fragments of my life were here and there, with no real rhyme or reason or design.

It was an amazing thing—and frankly, rather daunting—to face a future that had an almost unlimited number of options!

On one day I'd think, *Maybe I'll start another business.* But what type? Where? To provide what goods or services?

On another day I'd think, *Maybe I'll take up golf again—and see how many golf courses I can play in different states over the next year or so.* But for what real purpose? With whom?

On yet another day I'd think, *Maybe I'll volunteer . . .* But where? Doing what?

I knew that whatever I finally settled on doing, it would have to be something that challenged my *mind* and had me thinking, "How do I deal with this? How can I capitalize on that? What can I do to maximize my time, energy, and resources?" I think I'm a man in a never-ending quest for new puzzles to solve.

The first step for ME was to make some decisions about

where I wanted to live. I settled in Colorado, about an hour's drive from Buford.

I also spent some time thinking about what I didn't and don't want to do.

I don't have any desire to take long road trips or travel extensively. I've been there, done that. In the first full year after leaving Buford, I only took out my motorcycle twice—and one of those "excursions" was to ride around my new neighborhood, just to make sure the motorcycle was still in good working order.

I might, however, take a few short vacation trips to Europe. I have a strong interest in visiting Switzerland, which was my mother's birthplace.

And, I might want to visit Australia. When I was in the military, I traveled to Australia on leave, and I somewhat considered the offer that was being made by the Australian government at the time. The Aussies had a plan that would allow an American G.I. to locate there and be set up in either a rural or town setting. I've wondered periodically through the years what might have been my life if I had accepted the Australian offer. Would I have been living in the Outback running a little store? Who knows? But I hasten to add . . . that's not for me NOW.

I have had no interest in a repurchase of Buford. However, I did muse from time to time that I *might* entertain an offer to lease the store from the current owner, but only for a store operation in the summer months.

There were dozens of mights, maybes, and other possibilities to consider. I was really only sure of one thing: I refused to see the coming years as boring!

Destiny as a Wild West Ghost Town?

A CNN commentator quipped before the sale, "Buford? Without Don Sammons? It's unthinkable!"

He meant it as a joke, of course.

It may have been a prophecy.

In the first year after the sale, I only went back to Buford a few times. It was amazingly easy to walk away and not look back. I went to Cheyenne for concerts and to see friends, but on those trips, I never felt compelled to get on I-80 and drive west. I'm not sure why. Maybe I feared emotions might be dredged up that I didn't want to face.

I *was* surprised that the story—the *idea*—of Buford seemed to linger. The website, which I still have for Buford, still has lots of hits—currently, well over two-and-a-half million visits total.

Not long ago I heard from a childhood friend in St. Louis. We had lost touch with each other. She was surfing the Internet one evening, reading various articles that captured her attention, and she read about the sale of America's smallest town. She recognized my name, of course, and called me in excitement. She had known nothing of the sale of Buford, and frankly, didn't even know a place named Buford existed in Wyoming, but she was intrigued with the story. I suspect there are many more like her, regardless of whether I've ever met them in my past.

Williams & Williams went on to sell another "town" several months after the sale of Buford. They sold a place on I-90 in the area of the battlefield of the Little Big Horn conflict in South Dakota. The owner there had two buildings, one of which was a little museum. The property, however, didn't capture the same level of the media interest that the sale of Buford had.

In the year after the sale, Williams & Williams did a documentary on me, Buford, and the sale. They are using it as their prime promotional tool to show potential sellers what might be involved in an auction and how valuable it was to position a property for maximum "intrigue."

The story itself seems to have ongoing "sizzle"—perhaps a *unique* sizzle.

The Lessons of Buford Linger

I certainly have had ample time to reflect on a number of broader philosophical issues since the sale.

Through the years, a number of people asked me directly or indirectly, "Why you?" Why did Don Sammons, of all people, take on Buford and turn it into a national story?

As I indicated in the first chapter, the better question might be, "Why *not* me?"

Did anything uniquely qualify me for what happened at Buford?

The short answer is "no."

The guy who looks back at me out of my mirror is a man who was a kid like countless millions of kids in America. I loved the idea of the West, loved freedom to chart my own course in life, and loved things that weren't entirely "conventional." As I grew up, I continued to love those things. I wasn't inclined to compromise, or follow rules that made no sense to me.

That was true then . . . and now. It may have been

something embedded in the fact that I was an only child who was enrolled in a Catholic elementary school, but I'll leave that analysis to someone who feels a need to do it.

I greatly enjoyed interaction with people, and I enjoyed buying and selling (after I got the hang of it). In the course of running the Buford Trading Post I enjoyed making breakfast and lunch for travelers, and I enjoyed the marketing and advertising processes involved in enticing more and more travelers to stop by the store. I enjoyed being my own boss.

I wasn't afraid of working hard and taking a few risks.

I suspect there are literally millions of people in our nation—and countless millions around the world—who could honestly say, "I enjoy the same things Don Sammons liked. And I'm not afraid of hard work and taking risks."

So why don't more people end up moving to places like Buford, or doing what I did in turning a quick-stop gas station and store into a town. I think it comes down to two things: endurance and work.

Lots of work.

Lots of endurance.

It also takes a little ingenuity and a desire to have a *business.*

I learned at Buford *how* to be a businessman—and a good businessman.

The statistics say that most businesses fail during the first five years. That certainly could have happened to me. Several things bolstered my ability to succeed—factors I recognize now more than I recognized them then.

It helped to be the only game in town—so to speak. There was no competition for twenty-five miles in either direction.

It also helped that a business had been established in the location prior to my taking over ownership and management. That previous business had not thrived, necessarily, but it had survived.

It helped that I *wasn't* an experienced businessman, per se, in another location. I had no preconceived ideas about "the way to do things." I was open to experimenting, trying and failing (to a degree, of course), and to exploring new ideas and possibilities.

I had a career that I *could* have gone back to if the Buford Trading Post had been an utter failure. I didn't consider going back—frankly, I was too busy working to make the store a success. But I *could* have returned to the moving business.

I had a degree of flexibility when it came to the products I carried and the prices I charged.

And, I had a huge incentive to succeed and a track record of being willing to work hard. I had a son to raise and I needed to succeed at Buford, in part, to provide the stability I wanted for him.

I didn't think of myself as a rugged individualist in the early years of my life. I *do* see myself that way now. Buford, in many ways, made me a rugged individualist.

I think being an individualist is directly related to being willing to *change*—and that means being willing to take chances.

That's at the core of the "Wild West" way of life, in my opinion. People who settled the West were willing to take chances and to make dramatic changes in their lives. They took on the adventure of finding new places and developing new ways of surviving in those places.

Yes, I rode horses. I became very comfortable on the back of a horse.

Yes, I lived in a log house. I liked it, a lot.

Yes, I lived out on a ranch and learned how to be a rancher.

And yes, I took on ten acres along an interstate and developed it into a "town."

Who knows what *you* might do or become?

Start dreaming!

A New Population—And a New International Dream

One thing that never seemed *good* to me was that Buford, Wyoming, might have NO population. There's no sizzle in that! I had a deep longing that a new dream might take root there and I was delighted, therefore, when I began to hear rumblings that Nguyen and his associates were developing a plan for the revitalization of Buford. I was eager to see what they might do with the town.

And wow . . . what a plan it is turning out to be.

19

TAKING ON THE WORLD

WITH A CUP OF COFFEE IN MY HAND

For about a year after the Buford Trading Post was sold, nothing happened there. It was a "ghost town," of sorts. Nguyen uses the term, "lost town."

Then, one day about fifteen months after I turned the keys of the Trading Post over to the new owner, I heard from Nguyen. He and his associates began asking me to help them in a consulting role to get the store up and running, and to give them business insights that might help them as they took the store in a new direction.

At first, I said "no." I had no strong desire to return to the store and face the same routine and pressures. But the more I talked with Nguyen and his associates, the more I saw their vision for what they hoped to accomplish. It was a dream that meant doing business *from* Buford, not only *in* Buford. That was a new direction, with new possibilities and new challenges. I finally said "yes" to the challenge of growing Buford beyond Wyoming, beyond the United States, and on to the whole wide world.

In many ways, my involvement with Nguyen and the Vietnamese seemed to be a good conclusion for my own experience in Vietnam. Here was an opportunity for peace, for friendship, for a creative adventure, for capitalism.

I was especially pleased, of course, that the store was going to be functioning again and serving travelers on I-80. That was a validation, of sorts, that my work of three decades was worthy to be continued.

Halfway Around the World a Man Surfs the Internet . . .

At the time of the sale, I knew virtually nothing about Nguyen. He was a mystery to me. And his contact of me fifteen months after the sale became a great opportunity for me to hear HIS side of the story. I'm going to let him tell his story in his words . . .

A Lifetime Opportunity

by Pham Dinh Nguyen

In March of 2013 I was browsing the Internet one night and I saw a short article about a man who was preparing to sell his town—Buford, Wyoming, population 1—in an Internet auction. I was struck by several ideas, the first of which was that this was the smallest town in the United States. I looked at the photos of the property.

The more I read and looked, the more excited I became.

Since I was a boy, I have been a fan of the "cowboy movies." At that time, I didn't know exactly where Wyoming was, but I knew it was in the American "Wild West." The thought that a person could buy an entire town there was stunning.

In Vietnam, of course, there is no such idea that one person might own an entire town, or live in a place with "population 1." As difficult as it might be for an American to comprehend the idea, there is no concept of a place with such a wide-open landscape as

Wyoming, or for that matter, the concept that a person might ever be alone—doing anything, at any time, in any place. If a person has not lived in an Oriental city where there is no idea of aloneness or privacy, it is impossible to understand that I was pursuing an IDEA that did not seem connected to reality! Take that one step further . . . in Vietnam, there is no idea that a Vietnamese citizen might buy a population-one place in the United States of America. Own an American town? Impossible. And yet, here was the possibility!

I was struck with the realization: This is a once-in-a-lifetime opportunity! And I admit, I became completely absorbed with the idea. I knew I needed to try to buy this town . . . even though in March of 2013 it seemed like an impossible dream.

First, I knew I needed permission both to leave Vietnam and to enter the United States. The filing and approval of that paperwork usually takes several weeks, including a face-to-face interview with the U.S. Ambassador's staff in Saigon. It didn't seem possible. Nevertheless, I filed an "Emergency Request"—actually, it felt like an emergency. I had a very strong urgency that this was my future, my whole life ahead, and I needed to be in Wyoming for that auction.

I filed the request about nine in the morning and received an answer that very afternoon that I would be given an appointment. I called right away. At the interview I was asked three questions:

- Were you ever in Japan? Yes. I have a friend who is a physician and lives there. I visited him.
- How many days were you in Japan? About two weeks.
- Why do you want to go to the U.S.A.? I want to fly there to join the auction of Buford, the smallest town in the United States.

I have no idea why I was asked about Japan. And I have no idea why I wasn't asked a lot more questions about the auction or Buford.

There were no questions about whether I had the money to buy the town or what I intended to do with it. I'm glad I wasn't asked, because at least for the second question, I had no answer. What was important to me in that time was TRYING to buy the town. I was filled with the idea of WHAT I wanted to do—buy Buford—but I hadn't developed a WHY for doing so. I knew that would come if I won the auction.

I should also state that I had never been to an auction before I came to the United States to participate in one! There is no such thing in Vietnam. But again, I had seen auctions in movies. An auction seemed like an exciting event, although what I had seen in the movies—showing mostly auctions of artwork and artifacts in London or New York City—didn't turn out to be very similar to what happened in Buford!

My First Visit to the U.S.

The day after I got the necessary visa, I scheduled a flight to Cheyenne. I knew I needed to arrive a couple days early to meet with Rosie, the realtor-auctioneer who had been a referral from Williams & Williams, the company out of Oklahoma that was handling the auction. It is a legal requirement in Wyoming that a buyer have a designated realtor who is also an auctioneer, and Rosie and her husband have a company that does both.

So, I flew to the U.S. by way of Japan and then Dallas, Texas, and Denver, Colorado. One of my friends came with me—a man who had been to the United States many times, speaks English fluently, and understands the American culture very well. He was and is a friend, not a business associate. The truth is, none of my business associates knew WHY I was making the trip to the United States. I feared that if they knew I was heading halfway around the world to try to buy a town with a population of one, in a Wild West state in the U.S., they would think I had lost my mind. Who knows? They might have tried to have me committed for mental evaluation and kept me from traveling! I simply told them, "I'm going on business."

I gave this same reason to my wife. She loves me completely, but I knew that she would also question my sanity, and would probably try to stop me. And if she couldn't stop me, she definitely would worry—what if I tried and I didn't win? How could I succeed in a place I'd never been? How could I make it in such a big country where I didn't know anybody and was still learning the American culture? She would have worried about my safety, and probably my disappointment if I lost.

I did confide my idea and my travel plans to a few close friends, and they each advised me not to go, not to try. But I had to make my own decision. It was my dream and my idea.

So, there I was in Cheyenne, just a few weeks after I first heard about the sale of Buford, Wyoming. Rosie met me at the airport, and the next day, she drove me out to see Buford. I was amazed at the wide-open spaces—as far as I could see, there were no buildings, no people, and on a wintry day, not that much traffic on a major highway.

Rosie told me later that she was driving on a sheet of ice the last five miles before the Buford exit on I-80—she had tires that allowed her to do that, driving slowly, without slipping off the highway. I was too excited and interested in what I was seeing all around me to consider the driving conditions, and in the warm vehicle, I didn't realize how cold it was outside. When I stepped out of her vehicle at the Buford Trading Post I was hit with such a blast of cold wind that I could hardly keep breathing. They told me later that a blizzard was on its way. Rosie admitted after the auction that she had thought I would probably withdraw from the auction after I saw Buford—she couldn't imagine how a man who had only known tropical weather all his life could possibly want to buy such a cold place. For me, the wind and the cold made it all the more exciting! After all, nothing about the Wild West in the movies was easy. I also decided that late April and early May must be winter in Wyoming and that surely things would warm up.

The next day was the auction. We arrived early to get a good place. Several people from the media seemed surprised to see an Oriental face and they wanted to interview me. I declined their requests. I didn't really want to be known at that point, although it was exciting to have people from the major media networks want to talk to me.

The man leading the auction gave some information and some instructions. And then it began!

I was amazed at how fast it went! Suddenly we were at $500,000!

At that point, the contest seemed to narrow until it was only between two of us—we began bidding in increments of $10,000 . . . 640 . . . 650 . . . 660 . . . and so forth. Very quickly we were at 840 . . . 850 . . . 860, and I decided that when it was my turn to bid 890, I would skip that number and go directly to $900,000. I looked at the other man who was bidding—he was on the phone, apparently talking to the person he was representing at the auction—and he didn't respond. Finally, he looked at the auctioneer and waved his hand as if to say "no," and the auctioneer responded with "SOLD!" to the person who had bid $900,000.

I was that man!

Rosie turned to me and said, "You won! Congratulations, you are the new owner of Buford."

The entire auction was over in eleven minutes.

I had prepared for this moment for weeks, had traveled many hours to get to Cheyenne, and I had won a town in an auction in a matter of minutes. I was a little dazed but also excited in a way I had never experienced excitement before.

We went into the store to sign the necessary papers.

I was pleased with what I saw at the store, and at the house. I had no regrets.

And Now What to Do?

I won! And now, I had to figure out what to do with the town!

I am a very practical man. I am a businessman. It was one thing to have a dream and to win an auction, but now I needed to figure out how to make my "win" a profitable venture. The money for the purchase of Buford was not money from either of the companies in which I am involved. It was money loaned from friends, as well as my own money. I became intent on repaying those loans!

In Vietnam, I serve as the top executive in two companies.

I am the CEO of a company in Vietnam that is a major distribution company. We work with a number of international companies that export commodities to Vietnam, many of which are perishable. It is our role to receive those commodities and distribute them with a very fast turnaround. Some of the items are food staples that need to get to consumers as quickly as possible. We distribute throughout the Mekong Delta and Ho Chi Minh City, and some items go north as far as Hanoi. We deal with *imports*; we are not an exporting company.

I am also the chairman of PhinDeli Coffee. We have coffee plantations in the highlands of Vietnam, and I personally believe we process the best coffee in the world!

Phin means "filter" in Vietnamese. *Deli* is short for "delicious." Our filter coffee is just that: DELICIOUS. Vietnamese coffee is strong, but not bitter. Most coffees can be brewed to be very strong, but when they are made strong, they become bitter. Not PhinDeli.

Our coffee products are meant to be used by pouring hot water over the coffee in a filter container. The resulting beverage is usually blended thoroughly with sweetened condensed milk, and then served over ice. Now *this* is a product worth exporting . . . and the more I explored that idea with our leaders at PhinDeli Coffee, the more the purpose for Buford came into focus.

Buford would become the focal point for *worldwide* distribution of PhinDeli coffee. That is the path we began to pursue with all our energy.

We certainly intend to reach *all* of America, and then *all* of the world. Distributorships. Sales outlets. Direct consumer purchases. We will promote our coffee on Amazon and other Internet sites. But the starting point, and the "headquarters," will be Buford.

People have asked me if I am planning to turn Buford into a Vietnamese town. They seem to be imagining that I would want to create a major city around the store. Not at all! They seem surprised when I say that I want Buford to remain "population 1." That is its magic. That is what sets Buford apart from all other towns in the United States and in the world. Here is the world's smallest town—unimaginable—and that is the major reason it can have the biggest impact!

The world's #1 coffee from a town that's #1—population 1, that is.

The world's best coffee from the world's smallest town.

Without a doubt, EVERY person living in Buford enjoys PhinDeli coffee.

PhinDeli coffee will ride into the world—from the top of the world—the Wild West of the U.S.

Our coffee is truly a "dessert" product—the sweetened condensed milk makes it that way. And no coffee stands up to the sweetened milk like PhinDeli. I feel very confident about that.

Our PhinDeli mugs have been designed by one of Vietnam's finest designers. They will also be part of our marketing plan. Other products are going to be sought out and offered. Some may directly relate to coffee, others might not.

The Book into Vietnamese

I am proud to call Don Sammons my friend.

I have been to the United States several times now and Don has welcomed me into his home. We have had long hours of conversation. My wife has never met Don, but I have told her everything Don has said to me, and she feels as if she knows him and that he is her friend, too.

My children—five-year-old twins, a boy and a girl—are excited about their father traveling to and from the United States. My son is a huge fan of horses, so every trip to Cheyenne means a trip to the museum and the museum store there so I can bring him something related to horses or cowboys. He is very excited about *everything* related to horses, and in his imagination, he is one of the finest young cowboys in all Vietnam.

Don seems surprised that he is something of a celebrity in Vietnam—quite famous, actually. People are as interested in his story as I was in the beginning. I think he is an amazing man, very accomplished in many ways. It is something to be admired by all people that he developed Buford into the "town" it became. We are building on a great accomplishment.

We are bringing Don to Vietnam to tell more about his story and his book, *BUFORD One.* I have no doubt he will be very well received.

I certainly am aware that Don was a soldier for the United States during the conflict in Vietnam. Frankly, that was over before I was born. Many people have sad memories of that time, but the prevailing feeling in Vietnam today is not one of conflict. It is one of friendship. It is one of working together and creating together.

I think it is something tremendous that a former "enemy" can now be a great friend, and an ambassador for a brand created in Vietnam.

A Flashback Experience

When Nguyen first told me his story, I had a flashback moment or two to the time when I purchased the store thirty-two years ago. Nobody had thought my idea of owning the store was a good idea. The financial ledgers kept by the previous owner were in disarray and after a trusted friend and accountant looked at the paperwork I gave him, he said, "I'd distance myself from that as far as I could run."

In the end, it was my decision. I got a new accountant and bought the store.

I had ideas about what *might* be done. I had a dream. And that was what really mattered in the end.

I can relate to the fact that Nguyen is dream-driven. I am convinced that if a person has a strong dream, ideas about how to turn that dream into a reality will come!

The revitalization of Buford has been a reaffirmation to me that Buford is not only an historic place with nearly 150 years of name identity and activity . . . it is not only a convenience store with lots of bells and whistles associated with it . . . it is ultimately an IDEA.

Buford is an IDEA that stands for the pursuit and fulfillment of dreams!

It stands for:

- Wild West independence and accomplishment

- What one person can achieve over time, given hard work and diligence
- The pursuit of a risk that becomes a risk WORTH pursuing
- Uniqueness in marketing
- Diverse streams of profit

IDEAS sell products and services. Buford is living proof of that.

A Truly *GRAND* Reopening

The store reopened for business in June of 2013, but gradually. The boards were taken from the windows, the fuel tanks were filled, the house was opened and aired out.

The official reopening was the day after Labor Day in 2013. It was truly an amazing *grand* reopening.

A big white party tent was placed next to the store. PhinDeli coffee was served from there, and after the official ceremony, hosted by a leading Vietnamese actress, those in attendance were invited into the store, and each person was given a gift sack with several items, including coffee, a coffee filter, and a PhinDeli mug.

The people in the area rallied around the reopening in an amazing way. The sign people did an astounding job—they constructed large signs with steel frames to withstand the Wyoming winds in record-breaking time, even though they had a huge load of work already in their warehouse. The signs were up two days before the event and with very high quality.

Many people commented to me and to the new owners that they were glad the store had reopened—they said I-80 just wasn't the same without it. The Buford Trading Post had been there as long as any local resident could remember, and most people had taken its existence for granted. Once it was gone, there was no place to stop between Laramie and Cheyenne. Plus, something unique had disappeared from their lives.

A chef at a notable Cheyenne resort had come to Wyoming from Los Angeles, where he frequently made Oriental dishes. He was eager to step up to create a fabulous buffet menu featuring Vietnamese foods—from both fresh and fried spring rolls, to small Vietnamese style sandwiches, to skewered shrimp, and other dishes—each with an authentic sauce.

Nguyen and others on his team spoke. I spoke. People applauded, the ropes covering the tarp over the new sign were cut, and the large and colorful PhinDeli Town Store sign was revealed!

The crowd of residents and media representatives—radio, print, and television, and at least one movie producer—was friendly and happy to be there on a blue-sky, white-cloud, breezy day! Other reporters called in, a follow-up to the "sale" of the store eighteen months before. News of the reopening made *The New York Times*, *The New Yorker* magazine, and the *Los Angeles Times.*

It was a new start, and one that is clearly aimed at the WORLD.

Nguyen was accurate in saying that I find it difficult—a bit—to see myself in a celebrity role, especially in Vietnam.

But I'm starting to pack my bags. This could be a fun ride, and one that just might have good international-relations benefits for all involved.

I assure you, I am *not* bored.

For ongoing updates and to order additional copies of *BUFORD One*, go to the Buford website—www.bufordtradingpost.com—or check out Amazon.com.

For the Vietnamese translation of *BUFORD One*, you can also go to the PhinDeli website: www.phindeli.com.

Don Sammons

After an extended tour of overseas duty with the U.S. Army, more than fifteen years as a highly successful owner-operator for a major moving company, and experience as a rancher, Don Sammons purchased a convenience store and fuel station along I-80 between Cheyenne and Laramie, Wyoming. Over the next twenty years, he turned the Buford Trading Post into an internationally recognized TOWN. Buford, population 1, became world famous as America's smallest town. Sammons sold his town by means of an Internet auction in April 2012, and is presently consulting with the new owners of the property to help them reach their goals for turning Buford into an international distribution center for Vietnamese coffee products. Sammons is available for media interviews, motivational speaking, and business consulting at www.bufordtradingpost.com.

Made in the USA
Charleston, SC
07 December 2013